Laboratory Exercises for Electronic Devices
Eighth Edition

A Laboratory Manual to accompany *Electronic Devices*,
Eighth Edition, by Thomas L. Floyd

David M. Buchla
Steven Wetterling

PEARSON

Prentice
Hall

Upper Saddle River, New Jersey
Columbus, Ohio

Editorial Assistant: Lara Dimmick
Production Editor: Rex Davidson
Production Manager: Matt Ottenweller
Design Coordinator: Diane Ernsberger
Cover Designer: Linda Sorrells-Smith
Cover Photo: Getty Images
Director of Marketing: David Gesell
Marketing Manager: Jimmy Stephens
Marketing Assistant: Les Roberts

This book was printed and bound by Bind-Rite Graphics. The cover was printed by Phoenix Color Corp.

Pearson Prentice Hall™ is a trademark of Pearson Education, Inc.
Pearson® is a registered trademark of Pearson plc
Prentice Hall® is a registered trademark of Pearson Education, Inc.

Pearson Education Ltd.
Pearson Education Singapore Pte. Ltd.
Pearson Education Canada, Ltd.
Pearson Education—Japan

Pearson Education Australia Pty. Limited
Pearson Education North Asia Ltd.
Pearson Educación de Mexico, S.A. de C.V.
Pearson Education Malaysia Pte. Ltd.

10 9 8 7

ISBN-13: 978-0-13-242971-9
ISBN-10: 0-13-242971-3

Preface

This laboratory manual uses a new format that is carefully coordinated to the text *Electronic Devices*, Eighth Edition, by Thomas L. Floyd. The eighteen experiments correspond to the eighteen chapters in the text. All of the experiments are subdivided into two or three "parts". With one exception (Experiment 12-B), the parts for all experiments are completely independent of each other. The instructor can assign any or all parts of these experiments, and in any order. This format provides flexibility for the instructor depending on the schedule, laboratory time available, and course objectives.

In addition, Experiments 12 through 16 provide two options for experiments. These five experiments are divided into two major sections identified as A or B. The A experiments continue with the format of previous experiments; they are constructed with discrete components on standard protoboards as used in most electronic teaching laboratories. The A experiments have been tested in classrooms over many years and can be assigned in programs where traditional devices are emphasized.

Each B experiment has a similar format to the corresponding A experiment, but uses a programmable Analog Signal Processor (ASP) that is controlled by (free) Computer Aided Design (CAD) software from the Anadigm company (www.anadigm.com). These experiments support the Programmable Analog Design feature in the textbook. The B experiments are also subdivided into independent parts, but Experiment 12-B, Part 1, is a software tutorial and should be performed before any other B experiments. This is an excellent way to introduce the ASP technology because no other hardware is required other than a computer running the downloaded software. In addition to Experiment 12-B, the first 13 steps of Experiment 15-B, Part 2, are also tutorial in nature for the AnadigmFilter program. This is an amazing active filter design tool that is easy to learn and is included with the AnadigmDesigner2 (AD2) CAD software.

The ASP is part of a Programmable Analog Module (PAM) circuit board from the Servenger Company (www.servenger.com) that interfaces to a personal computer. The PAM is controlled by the AD2 CAD software from the Anadigm company website. Except for Experiment 12-B, Part 1, it is assumed that the PAM is connected to the PC and AnadigmDesigner2 is running. Experiment 16-B, Part 3, also requires a spreadsheet program such as Microsoft® Excel®. The PAM is described in detail in the Quick Start Guide (Appendix B). Instructors may choose to mix A and B experiments with no loss in continuity, depending on course objectives and time. We recommend that Experiment 12-B, Part 1, be assigned if you want students to have an introduction to the ASP without requiring a hardware purchase.

A new text feature is the Application Activity (AA) at the end of most chapters. All of the AAs have a laboratory exercise using a similar circuit that is sometimes simplified to make laboratory time as efficient as possible. An icon identifies the AA exercise in the lab manual.

The format for each experiment is:

- **Introduction**: A brief discussion about the experiment and comments about each of the independent parts that follow.
- **Reading:** Reading assignment in the Floyd text related to the experiment.
- **Key Objectives:** A statement specific to each part of the experiment of what the student should be able to do.
- **Components Needed:** A list of components and small items required for each part but not including the equipment found at a typical lab station. Particular care has been exercised to select materials that are readily available and reusable, keeping cost at a minimum.
- **Parts:** There are two or three independent parts to each experiment. Needed tables, graphs, and figures are positioned close to the first referenced location to avoid confusion. Step numbering starts fresh with each part, but figures and tables are numbered sequentially for the entire experiment to avoid multiple figures with the same number.
  - **Conclusion:** At the end of each part, space is provided for a written conclusion.
  - **Questions:** Each part includes several questions that require the student to draw upon the laboratory work and check his or her understanding of the concepts. Troubleshooting questions are frequently presented.
- **Multisim Simulation**: At the end of each A experiment (except #1 and #18), one or more circuits are simulated in Multisim. Several new Multisim problems have faults for troubleshooting practice. These are identified with the word *fault* in the file name. Other files, with *nf* as the suffix of the filename, include measurement exercises using instruments such as the Bode Plotter and the Spectrum Analyzer. A special icon is shown with all figures that are related to the Multisim simulation. Multisim files are found on the Floyd website at **www.prenhall.com/floyd**. Click on the *Electronic Devices* book.

Multisim

Microsoft PowerPoint® slides are available at no cost to instructors for all experiments. The slides reinforce the experiments with troubleshooting questions and a related problem. Contact your Pearson/Prentice Hall representative to obtain a CD with the slides.

Each laboratory station should contain a dual-variable regulated power supply, a function generator, a multimeter, and a dual-channel oscilloscope. A list of all required materials is given in Appendix A along with information on acquiring the PAM. Components are also available as a kit from *Electronix Express* as kit #32DBEDFL08 (www.elexp.com).

We have enjoyed the close collaboration with Tom Floyd on this manual. We also would like to thank Rex Davidson at Prentice Hall and Lois Porter, who copyedited the manuscript and made many excellent suggestions along the way.

David Buchla and Steven Wetterling

D.B.: To my granddaughter - Rachel Kaitlyn Holman

S.W.: To my son - Galen Holt Wetterling

Contents

Introduction

Safety in the Laboratory

The experiments in this lab book are designed for low voltages to minimize electric shock hazard; however, one should never assume that electric circuits are safe. A current of a few milliamps through the body can be lethal. In addition, electronic laboratories often contain other hazards such as chemicals, power tools, and hot soldering irons. For your safety, you should review laboratory safety rules before beginning a course in electronics. In particular, you should:

1. Avoid contact with *any* voltage source. Turn off power before working on circuits.
2. Remove watches, jewelry, rings, and so forth before working on circuits – even those circuits with low voltages – as burns can occur.
3. Know the location of the emergency power-off switch.
4. Never work alone in the laboratory.
5. Keep a neat work area and handle tools properly. Wear safety goggles or gloves when required.
6. Ensure that line cords are in good condition and grounding pins are not missing or bent. Do not defeat the three-wire ground system in order to make "floating" measurements.
7. Check that transformers and instruments that are plugged into utility lines are properly fused and have no exposed wiring.
8. If you are not certain about a procedure, check with your instructor before you begin.
9. Report any unsafe condition to your instructor.
10. Be aware of and follow laboratory rules.

Preparing for Laboratory Work

The purpose of experimental work is to help you gain a better understanding of the principles of electronics and to give you experience with instruments and methods used by technicians and electronic engineers. You should begin each experiment with a clear idea of the purpose of the experiment and the theory behind the experiment. Each experiment requires you to use electronic instruments to measure various quantities. The measured data will be recorded and you will need to interpret the measurements and draw conclusions about your work. The ability to measure, interpret, and communicate results is basic to electronic work.

Preparation before coming to the laboratory is an important part of experimental work. You should prepare for every experiment by reading it over and checking the reference reading. This prelab preparation will enable you to work efficiently in the laboratory and enhance the value of the laboratory time.

This laboratory manual is designed to help you measure and record data as efficiently as possible. Techniques for using instruments are described in many experiments. Data tables are prepared and properly labeled to facilitate the recording of data. Plot space is provided to graph the voltage waveforms that you will see. You will need to interpret and discuss the results in the *Conclusion* section. The *Conclusion* to an experiment is a concise statement of your key findings from the experiment. Be careful of generalizations that are not supported by the data. The conclusion should be a specific statement that includes important findings that directly relate to the objectives of the experiment. For example, if the objective of the experiment is to measure $V_{GS(off)}$ and I_{DSS} for a JFET (as in Experiment 8), the conclusion should indicate these values as determined in the experiment. You should include a statement about experimental error and a comparison to theory.

Using the Programmable Analog Module

The "B" series of activities in Experiments 12, 13, 14, 15 and 16 introduce the student to the concept of computer-controlled analog signal processing using the Programmable Analog Module from the Servenger company (www.servenger.com) and the AnadigmDesigner2 CAD software from the Anadigm company (www.anadigm.com).

The Analog Signal Processor (ASP) IC is at the center of the Programmable Analog Module (PAM). It is a versatile IC that can be configured to perform a variety of analog signal processing tasks using the AnadigmDesigner2 program. For example, the ASP can be configured to add two signals together, find the difference between two signals, perform integration, differentiation, and filtering or create periodic waves such as sine waves, square waves or complex waves. Although the ASP is a clock-driven signal processor as will be explained in detail in the Experiments, it is *not* a digital signal processor. The signals inside the Anadigm ASP really are analog signals, and all the methods and mathematics of analog signal processing that you are studying apply.

You will find these experiments to be very educational and quite fun as you learn to use the CAD tools and the PAM. Once you become skillful, the time to design a new circuit, download it into the PAM, and then measure it on your oscilloscope will be just a few minutes.

The Laboratory Notebook

Your instructor may assign a formal laboratory report for you to write. A suggested format for formal reports is as follows:

1. *Title and date.*
2. *Purpose:* Give a statement of what you intend to determine.
3. *Equipment and materials:* Include a list of equipment model and serial numbers to allow retracing if a defective or an out-of-calibration piece of equipment is used.
4. *Procedure:* Give a brief description of what you did and what measurements you made. A diagram or schematic is often useful.
5. *Data:* Tabulate raw (unprocessed) data; data may be presented in graph form.
6. *Sample calculations:* Give the formulas that you applied to the raw data to transform it to processed data.
7. *Conclusion:* The conclusion is a specific statement supported by the experimental data. It should relate to the objectives for the experiment. For example, if the purpose of the experiment is to determine the frequency response of a filter, the conclusion should describe the frequency response or contain a reference to an illustration of the response.

Graphing

A graph is a pictorial representation of data that enables you to see the effect of one variable on another. Graphs are widely used in experimental work to present information because they enable the reader to discern variations in magnitude, slope, and direction between two quantities. In this manual, you will graph data in many experiments. You should be aware of the following terms that are used with graphs:

abscissa: the horizontal or *x*-axis of a graph. Normally the independent variable is plotted along the abscissa.

dependent variable: a quantity that is influenced by changes in another quantity (the independent variable).

graph: a pictorial representation of a set of data constructed on a set of coordinates that are drawn at right angles to each other. The graph illustrates one variable's effect on another.

independent variable: the quantity that the experimenter has control over.

ordinate: the vertical or *y*-axis of a graph. Normally the dependent variable is plotted along the ordinate.

scale: the value of each division along the *x*- or *y*- axis. In a linear scale, each division has equal weight. In a logarithmic scale, each division represents the same percentage change in the variable.

The following steps will guide you in preparing a graph:

1. Determine the type of scale that will be used. A linear scale is the most frequently used and will be discussed here. Choose a scale factor that enables all of the data to be plotted on the graph without being cramped. The most common scales are 1, 2, 5, or 10 units per division. Start both axes from 0 unless the data covers less than half of the length of the coordinate.
2. Number the *major* divisions along each axis. Do not number each small division as it will make the graph appear cluttered. Each division must have equal weight. *Note*: The experimental data is <u>not</u> used to number the divisions.
3. Label each axis to indicate the quantity being measured and the measurement units. Usually, the measurement units are given in parentheses.
4. Plot the data points with a small dot with a small circle around each point. If additional sets of data are plotted, use other distinctive symbols (such as triangles) to identify each set.
5. Draw a smooth line that represents the data trend. It is normal practice to consider data points but to ignore minor variations due to experimental errors. (*Exception*: calibration curves and other discontinuous data are connected "dot-to-dot".)
6. Title the graph to indicate what the graph represents. The completed graph should be self-explanatory.

The Oscilloscope

The oscilloscope is a fundamental laboratory instrument. An oscilloscope draws a graph of voltage versus time allowing you to visualize the circuit action. Most scopes today are digital based but are conceptually the same as the original analog cathode ray tube (CRT) oscilloscopes. The variety of controls on the front panel of a scope may seem confusing, but you can understand them better by keeping in mind the four fundamental sections that are basic to all scopes:

- The input signal is connected to the **vertical** section which can be set to attenuate or amplify the input signal to provide the proper voltage level.
- The **trigger** section samples the input waveform and sends a synchronizing trigger signal at the proper time to the horizontal section. The trigger occurs at the same relative time to superimpose each succeeding trace on the previous trace. This action causes multiple repetitions of the signal to appear to stand still on the display allowing you to examine it.
- The **horizontal** section is the time base for the graph. It controls the rate the voltage samples are taken on a digital scope or the rate the beam moves left to right on an analog scope.
- The **display** section has controls for adjusting how signals are viewed.

Digital scopes also provide a wide variety of measurement options that are specific to the particular vendor and model of scope; examples: voltage amplitude, relative time between events, averaging, peak detection, saving and retrieving previous waveforms, etc.

In your first laboratory session carefully identify the oscilloscope that you will be using. Pay attention to the various electrical connections, controls, and the measurement and display options provided. The details are explained in the operator's manual for that particular scope. Study it carefully. If the manual is not available at your workspace, then try the manufacturer's website or other resources on the web.

Oscilloscope Probes

Signals should always be coupled into the oscilloscope through a scope probe, which is normally supplied by the oscilloscope manufacturer. A probe is used to pick off the signal and couple it to the oscilloscope. Probes act to reduce the loading effect on circuits and to extend the frequency response of the measurement. Probes come equipped with a short ground lead with a grabber or alligator clip to make the electrical connection to the ground of the test circuit. Be careful that the ground lead does not accidentally touch an electrically "live" part of the circuit you are measuring. It should always be connected to a solid point such as a ground pin on the circuit board. The PAM units provide two ground terminals for this purpose.

To know that your oscilloscope and probes are making accurate measurements, begin every lab with a quick check of the scope probe compensation for each channel. Probe compensation is done by turning the adjustment slot seen in the little hole in the housing at the base of the scope probe nearest to the connection into the oscilloscope while connected to the "Calibrator" signal provided for this purpose. A correctly compensated scope probe shows a square-cornered, flat-topped square wave. Always repeat the adjustment when you move a probe to another channel or to another scope.

Experiment 1 Introduction to Semiconductors

Semiconductor diodes are created by the joining together of *p*- and *n*-materials at the time of manufacturing, creating a one-way valve for current. The region near the joined area is called the *depletion region*. If an external voltage is applied to the diode, it is said to be biased. Reverse-bias is the condition where a positive voltage is connected to the *n*-region and a negative voltage is applied to the *p*-region. This creates a widening of the depletion region and effectively reduces the current to zero. Forward-bias is the condition where a positive voltage is connected to the *p*-region and a negative voltage is applied to the *n*-region. This creates a narrower depletion region and allows charge carriers to cross, thus enabling current.

In Part 1 of this experiment, you will learn a simple diode test with a meter and measure the diode characteristic of a semiconductor diode. In Part 2, this work is extended to look at the characteristics with an oscilloscope and to plot other types of diodes.

Reading

Floyd, *Electronic Devices*, Eighth Edition, Chapter 1

Key Objectives

Part 1: Measure the forward and reverse characteristics of a signal diode and observe the relationship between current and voltage on a semilog plot.

Part 2: Plot the forward and reverse characteristics of a signal diode, a zener diode, and an LED using the oscilloscope and compare the forward ac resistance of each.

Components Needed

Part 1: The Diode Characteristic Curve

Resistors: one 330 Ω, one 1.0 MΩ
One signal diode, 1N914

Part 2: Plotting Diode Curves with an Oscilloscope

One 1.0 kΩ resistor
One 12.6 V ac transformer with fused line cord
One signal diode 1N914
Two LEDs, one red, one green
One zener diode, 1N4733A

Part 1: The Diode Characteristic Curve
Diode Test with a Meter

1. A quick diode test is useful if you are not sure if a diode is good. Digital meters usually have a *diode test position*, which allows the meter to provide bias voltage for the diode. In this configuration, a diode is placed between the leads. It will be forward-biased in one direction and reverse-biased in the other. Generally, you should read a small voltage in one direction (silicon diodes will read about 0.6 V) and an overload (*OL*) or an open reading in the other. Check the meter you are using and test a 1N914 diode using this method or one described by the manufacturer of your meter. If you do not have a diode test position on your meter, you can use an ohmmeter to test the forward and reverse resistance; however, this method is not as reliable as the first test as not all meters will read the same.

    Describe the meter test you performed:

    _____

    _____

2. Measure and record the values of the resistors listed in Table 1-1. It is a good idea to always measure resistors in experiments to help in understanding results later and to ensure that the resistors are the correct values.

Table 1-1

Component	Listed Value	Measured Value
R_1	330 Ω	
R_2	1.0 MΩ	

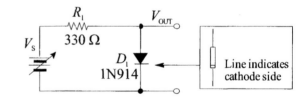

Figure 1-1

3. Construct the forward-biased circuit shown in Figure 1-1. The line on the diode indicates the cathode side of the diode (with forward bias, this is the negative side). The power supply represents the source voltage (V_S). Set it for zero volts (or the lowest voltage you can obtain).

4. Monitor the forward voltage drop, V_F, <u>across the diode</u>. Slowly increase V_S to establish 0.45 V across the diode. With this voltage across the diode, measure the voltage across the resistor, V_{R1}, and record it in Table 1-2.

Table 1-2

V_F	V_{R1} (measured)	I_F (computed)
0.45 V		
0.50 V		
0.55 V		
0.60 V		
0.65 V		
0.70 V		
0.75 V		

5. The diode forward current, I_F, can be found by applying Ohm's law to R_1. Compute I_F and enter the computed current in Table 1-2.

6. Repeat steps 4 and 5 for each voltage listed in Table 1-2.

7. The data in this step will be accurate only if your voltmeter has high input impedance. You can test the impedance of your meter by measuring the power supply voltage through a series 1:0 MΩ resistor. If the meter reads the supply voltage accurately, it has high input impedance. Connect the reverse-biased circuit shown in Figure 1-2. Set the power supply to each voltage listed in Table 1-3. Measure and record the voltage across R_2 for each voltage. Apply Ohm's law to compute the reverse current in each case. Enter the computed current in Table 1-3.

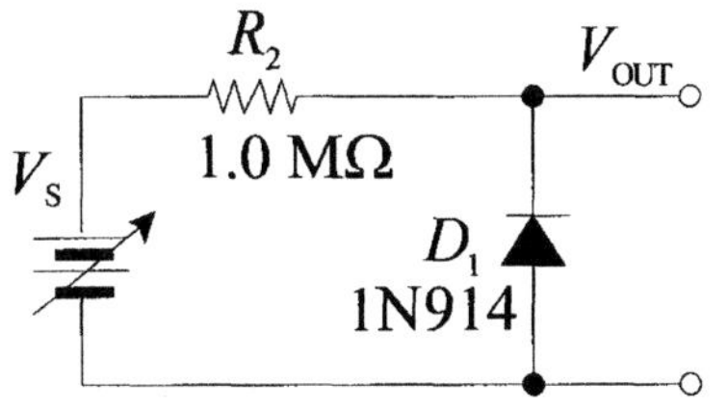

Figure 1-2

Table 1-3

V_S (measured)	V_{R2} (measured)	I_R (computed)
5.0 V		
10.0 V		
15.0 V		

8. Graph the forward- and reverse-biased diode curves on Plot 1-1, a linear plot. The different voltage scale factors for the forward and reverse curves are chosen to allow the data to cover more of the graph. You need to choose an appropriate current scale factor which will put the largest current recorded near the top of the graph. You should also add a title to your graph.

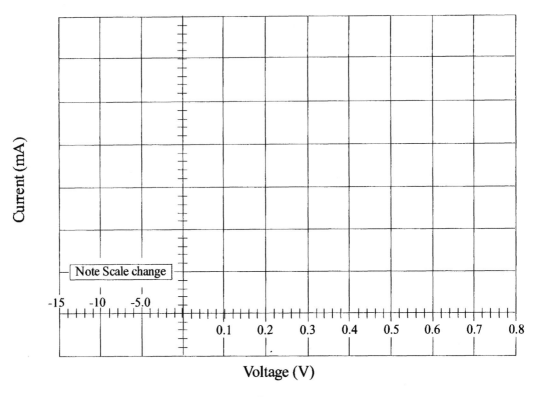

Plot 1-1

9. Plot the same forward-biased data from Table 1-2 in a semilog plot given as Plot 1-2 on the next page. The scales are already set up but you need to add a title to the plot. A semilog plot has a logarithmic scale on one axis and a linear scale on the other axis. What conclusion can you make from this plot?

_____

Conclusion: Part 1

Questions: Part 1

1. Assume you measure a power supply voltage through a series 1 MΩ resistor and observe that it drops by ½ its former value when no resistor is in series. What is the input impedance of the meter? Explain your answer.

2. Why is a large resistor used in series with the reverse-biased diode in step 7?

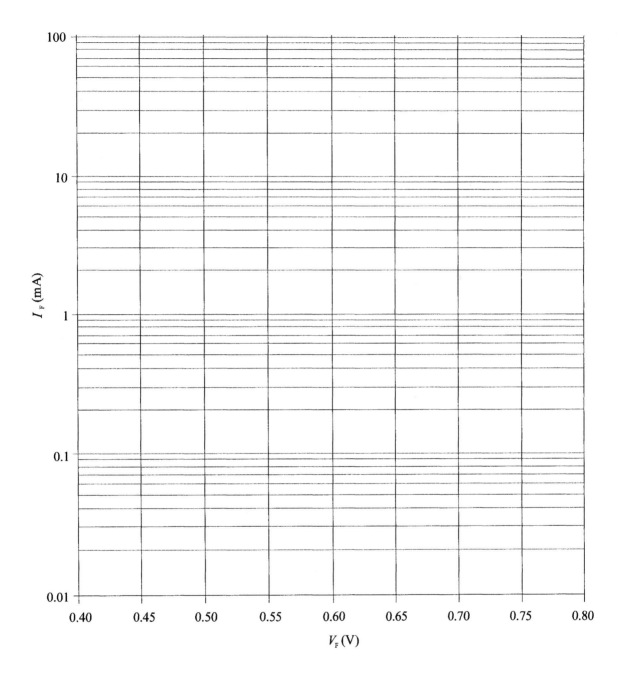

Plot 1-2

Part 2: Plotting Diode Curves with an Oscilloscope

Safety note: You are connecting a 12.6 VAC transformer into the ac utility voltage. Be certain the line cord is insulated and fused. At no time should you touch the circuit when it is connected. You will make no connections or measurements on the primary side of the transformer.

1. You can plot the characteristic curve for any diode by connecting the circuit shown in Figure 1-3. Notice that the transformer does not have a ground connected at any point. The scope is in the X-Y mode. Start with a signal diode (1N914 or equivalent). Channel 1 senses the voltage drop across the diode; Channel 2 shows a signal that is proportional to the current. Because the Channel 2 signal is actually the voltage across a 1.0 kΩ resistor, the V/div control can be converted to a mA/div control if you divide the V/div setting by 1.0 kΩ. Invert Channel 2 to display the signal in the proper orientation (if you cannot invert channel 2, you can still observe the response but it is upside-down).

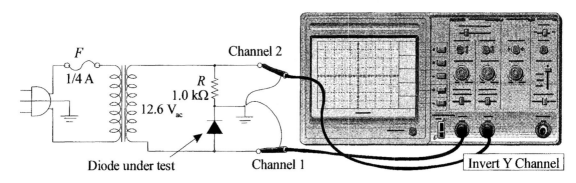

Figure 1-3

2. A good starting point for reading current and voltage is to center the display. Select the ground position on Channel 1 and center the vertical line with the Channel 1 position control. After it is centered, select dc coupling and repeat the procedure with Channel 2, this time centering the horizontal line with the channel 2 position control. The current and voltage can now be read directly from the display, measured from the center of the screen. Observe the dynamic resistance in the forward-biased region.

3. (*optional*) Test the effect of heat on the diode. If you have a hot soldering iron, bring it near (but not touching) the diode while observing the characteristic curve on the scope.
    Observations:

6

4. Turn off the power and replace the signal diode with a red LED. Reapply power. Measure the threshold voltage and the dynamic resistance of the LED in the forward-biased region.

    Observations:

    _____

5. Repeat step 4 with a green LED.

    Observations:

    _____

6. Repeat step 4 but use a 5 V zener diode (1N4733A or equivalent) instead of an LED. Measure the zener breakdown voltage.

    Observations:

    _____

Conclusion: Part 2

Questions: Part 2

1. Explain how to determine the dynamic resistance of a diode.

2. Could the setup in this Part be used to measure the characteristic curve of a resistor? Explain your answer.

Experiment 2 Diode Applications

In analyzing diode circuits, it is useful to use models of behavior as discussed in the text. In this experiment, the practical model is applied to all circuits. Recall that the practical model includes a 0.7 V drop when the diode is forward biased and an open circuit when the diode is reverse biased. Other effects can be ignored for most analysis work and in this experiment.

In Part 1, you will test three rectifier circuits. The first to be tested is the half-wave rectifier followed by the full wave rectifier. The third rectifier circuit you will test is a bridge circuit similar to the Application Activity in the text. The values are different in order to keep power down but still illustrate the concepts. Rectifier circuits are particularly important because they are a fundamental component in dc power supplies; they are introduced here and covered in more detail in Chapter 17 of the text and in Experiment 17. It is important to wire the circuit with **power off** and to pay particular attention to circuit grounds. An incorrectly wired circuit will blow the fuse!

Part 2 focuses on diode *limiting* circuits and Part 3 focuses on diode *clamping* circuits. Limiting circuits (also called *clipping circuits*) are used to prevent a waveform from exceeding some particular limits, either negative or positive. Clamping circuits are used to shift the dc level of a waveform, useful in signal processing and communication circuits.

Reading

Floyd, *Electronic Devices*, Eighth Edition, Chapter 2

Key Objectives

Part 1: Construct half-wave, full-wave, and bridge rectifier circuits, compare the input and output voltage for each, and observe the effect of filtering.

Part 2: Measure the clipping level and effect on the output waveform with various changes to a biased clipping circuit.

Part 3: Predict and measure the effect of dc bias voltage on a clamping circuit.

Components Needed

Part 1: Diode Rectifiers

Resistors: two 2.2 kΩ resistors
One 12.6 V ac center-tapped transformer with fused line cord
Four rectifier diodes 1N4001
One 100 µF capacitor

Part 2: Diode Clipping Circuits

Resistors: one 1.0 kΩ, two 10 kΩ, one 100 kΩ
One signal diode: 1N914

Part 3: Diode Clamping Circuits
One 100 kΩ resistor
One 100 µF capacitor
One signal diode: 1N914

Part 1: Diode Rectifiers

Safety note: You are connecting a 12.6 VAC transformer into the ac utility voltage. Be certain the line cord is insulated and fused. At no time should you touch the circuit when it is connected. You will make no connections or measurements on the primary side of the transformer.

Half-Wave Rectifier

1. Connect the half-wave rectifier circuit shown in Figure 2-1 using a 12.6 V ac transformer with a fuse. Notice that the transformer center tap is not connected. The line on the diode indicates the cathode side (the negative side when forward-biased). Compute the peak output voltage ($V_{out(p)}$) and enter the computed value in Table 2-1 (3rd column). Connect the oscilloscope so that channel 1 is across the transformer secondary and channel 2 is across the output (load) resistor. The oscilloscope should be set for LINE triggering as the waveforms to be viewed in this experiment are synchronized with the ac line voltage. View the input voltage, V_{sec}, and output voltage, V_{OUT}, waveforms for this circuit and sketch them on Plot 2-1. Label voltage and time on your sketch and add a title to the plot.

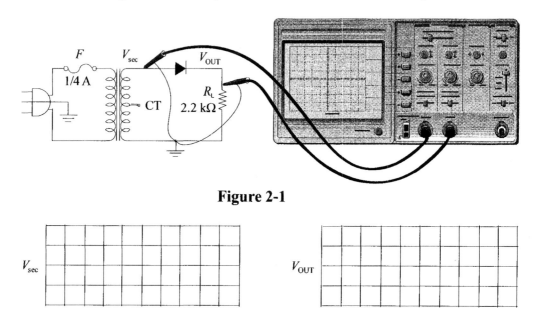

Figure 2-1

V_{sec}

V_{OUT}

Plot 2-1

2. Measure the secondary rms voltage and the output peak voltage. Remember to convert the oscilloscope reading to rms voltage. Record the data in Table 2-1.

Table 2-1 Half-wave rectifier.

Without Filter Capacitor				With Filter Capacitor		
Computed	Measured	Computed	Measured	Measured		
$V_{sec(rms)}$	$V_{sec(rms)}$	$V_{out(p)}$	$V_{out(p)}$	$V_{OUT(DC)}$	$V_{r(pp)}$	Ripple Frequency
12.6 V ac						

3. The output isn't very useful as a dc voltage source because of the pulsating output. Connect a 100 μF filter capacitor in parallel with the load resistor (R_L). Check the polarity of the capacitor – the negative side (indicated with a line) goes toward ground. Measure the dc load voltage, $V_{OUT(DC)}$, and the peak-to-peak ripple voltage, $V_{r(pp)}$, in the output. To measure the small ac ripple voltage, switch the oscilloscope to AC COUPLING. This allows you to magnify the ripple voltage without including the much larger dc level. Measure the ripple frequency. The ripple frequency is the frequency at which the waveform repeats. Record all data in Table 2-1.

Full-Wave Rectifier
4. Disconnect power and connect the full-wave rectifier circuit shown in Figure 2-2. Notice that the ground for the circuit has changed. Check your circuit carefully before applying power. Compute the peak output voltage ($V_{out(p)}$) and enter the computed value in Table 2-2 (3rd column). Then apply power and view the V_{sec} and V_{OUT} waveforms. Sketch the observed waveforms on Plot 2-2.

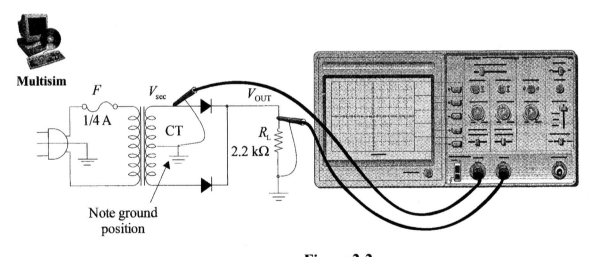

Figure 2-2

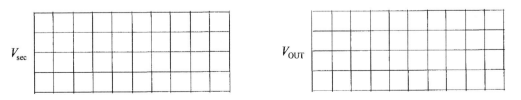

Plot 2-2

11

Table 2-2 Full-wave rectifier circuit.

Without Filter Capacitor				With Filter Capacitor		
Computed	Measured	Computed	Measured	Measured		
$V_{sec(rms)}$	$V_{sec(rms)}$	$V_{out(p)}$	$V_{out(p)}$	$V_{OUT(DC)}$	$V_{r(pp)}$	Ripple Frequency
6.3 V ac						

5. Measure $V_{sec(rms)}$ and the peak output voltage ($V_{out(p)}$) without a filter capacitor. Record the data in Table 2-2.

6. Now add a 100 µF capacitor in parallel with the load resistor. Measure $V_{OUT(DC)}$, the peak-to-peak ripple voltage, $V_{r(pp)}$, and the ripple frequency as before. Record the data in Table 2-2.

7. Connect a second 2.2 kΩ load resistor in parallel with R_L in the full-wave circuit in Figure 2-2. This will have the effect of doubling the load current. The filter capacitor should be left in parallel also. Measure the ripple voltage. What can you conclude about the effect of additional load current on the ripple voltage?

Application Activity

Bridge Rectifier

8. Disconnect power and change the circuit to the bridge rectifier circuit shown in Figure 2-3. In this configuration, the center tap of the transformer is not used and notice that <u>no</u> terminal of the transformer secondary is at ground potential. Because the input voltage to the bridge, V_{sec}, is not referenced to ground, you cannot use the oscilloscope to view both the secondary voltage and the output voltage at the same time. Check your circuit carefully before applying power. Compute the expected peak output voltage. Then apply power and use a voltmeter to measure $V_{sec(rms)}$. Use the oscilloscope to measure the peak output voltage ($V_{out(p)}$) without a filter capacitor. Record the data in Table 2-3.

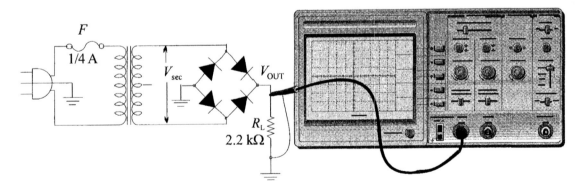

Figure 2-3

12

Table 2-3 Bridge rectifier circuit.

Without Filter Capacitor				With Filter Capacitor		
Computed	Measured	Computed	Measured	Measured		
$V_{sec(rms)}$	$V_{sec(rms)}$	$V_{out(p)}$	$V_{out(p)}$	$V_{OUT(DC)}$	$V_{r(pp)}$	Ripple Frequency
12.6 V ac						

9. Connect the 100 μF capacitor in parallel with the load resistor. Measure $V_{OUT(DC)}$, the peak-to-peak ripple voltage, and the ripple frequency as before. Record the data in Table 2-3.

10. Simulate an open diode in the bridge by removing one diode from the circuit. What happens to the output voltage? To the ripple voltage? To the ripple frequency?

Conclusion: Part 1

Questions: Part 1

1. In step 4, you moved the ground reference to the center tap of the transformer. If you wanted to look at the voltage across the entire secondary, you would need to connect the oscilloscope as shown in Figure 2-4 and add the two channels. Why is it necessary to use two channels to view the entire secondary voltage?

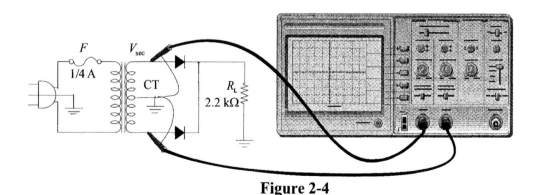

Figure 2-4

2. How can you measure the ripple frequency to determine if a diode were open in a bridge rectifier circuit?

13

Part 2: Diode Clipping Circuits

1. Connect the circuit shown in Figure 2-5. Connect the signal generator to the circuit and set it for a 6.0 V_{pp} sine wave at a frequency of 1.0 kHz with no dc offset. Observe the input and output waveforms on the oscilloscope by connecting it as shown. Notice that R_2 and R_L form a voltage divider, causing the load voltage to be a little less than the source voltage. R_1 will provide a dc return path in case the signal generator is capacitively coupled.

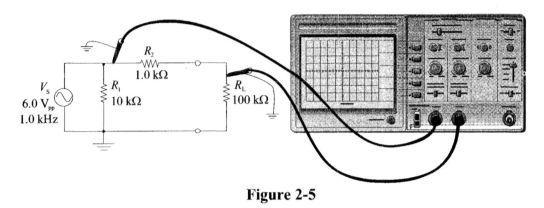

Figure 2-5

Observations:

_____

2. Add the 1N914 signal diode to the circuit as shown in Figure 2-6. Look carefully at the output waveform. Notice the zero volt level. Sketch the input and output waveforms in Plot 2-3. Then measure[1] the waveform across R_2. You can do this by viewing the difference between channel 1 and channel 2 (Ch 1 – Ch 2). Sketch the waveform on the third section of Plot 2-3.

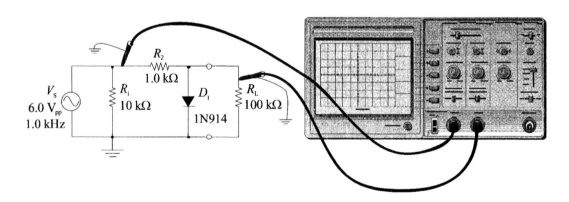

Figure 2-6

_____

[1] On the oscilloscope, you can view the difference between the two channels by placing the probes on both sides of R_2. The channels are set to the same vertical sensitivity (VOLTS/DIV). For analog scopes, you can select the ADD function and INVERT Ch-2. Check with your operator's manual if these functions are not available.

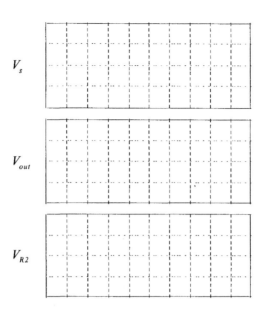

V_s

V_{out}

V_{R2}

Plot 2-3

3. Change the load resistor to 10 kΩ. Observe the effect of the smaller load resistor. Observations:

    _____

4. Restore the original 100 kΩ load resistor. Remove the cathode of the diode from ground and connect the cathode to the DC power supply as shown in Figure 2-7. Start at +2.0 V, and then vary the voltage from the supply and describe the results.

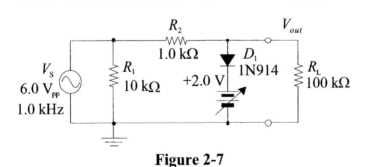

Figure 2-7

5. Reverse the diode in the circuit of Figure 2-7. Vary the dc voltage and describe the results.

    _____

15

6. Replace the positive power supply with a negative supply. Again, vary the dc voltage and describe the results.

Conclusion: Part 2

Questions: Part 2
1. In the circuit of Figure 2-6, did the clipping level change when the load was changed?

2. Given the waveform across the load and the source waveform, how could you use Kirchhoff's voltage law to predict the waveform across R_2 in Figure 2-6?

Part 3: Diode Clamping Circuits
1. Connect the clamping circuit shown in Figure 2-8. Couple the oscilloscope with dc coupling and observe the output voltage. Vary the input voltage.
    Observations:

Multisim

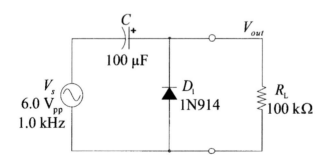

Figure 2-8

2. Add a dc voltage to the diode by connecting the power supply as shown in Figure 2-9. Sketch the output waveform on Plot 2-4. Show the dc level on your sketch.

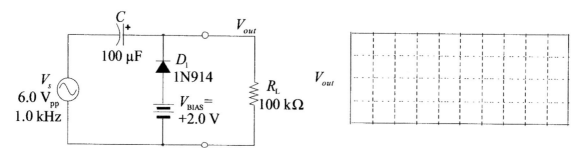

Figure 2-9   **Plot 2-4**

3. Reverse the diode and the capacitor. Then replace the positive dc voltage with a negative dc source and observe the output.
   Observations:

_____

Conclusion: Part 3

Questions: Part 3
1. When the diode and dc voltage were reversed in step 3, why was it necessary to reverse the capacitor also?

2. For the circuit of Figure 2-8, what change in the output would you expect if the load resistor were 10 kΩ?

Multisim Simulation

Multisim

Multisim files for the lab manual are on the website www.prenhall.com/floyd. The Multisim file Experiment 02_Full wave-nf is essentially the same as the circuit in Figure 2-2. (The nf suffix means there is *no fault*.) Open the file and compare your result to the computer simulation. The Multisim file Experiment 02_Clamping-nf is similar to the clamping circuit in Figure 2-8. Based on your knowledge of clamping circuits, predict the output for each circuit. Then test each circuit by connecting the oscilloscope and observing the waveform.

Experiment 3 Special-Purpose Diodes

Special-purpose diodes include the zener diode, the varactor diode, optical diodes (LED and photocell) and others. These diodes are investigated in this experiment.

In Part 1, you will observe the zener diode's *V-I* characteristic curve on an oscilloscope. You will then test the zener as a regulator using two circuits. In the first circuit, you will test the effect of a varying voltage and in the second circuit, you will test the effect of a varying load. In steps 10 and 11, you will test an IC regulator that uses an internal zener reference in a simple power supply like the one in the Application Activity in the text. As in Experiment 2, it is important to wire the circuit with **power off** and to pay particular attention to circuit grounds. An incorrectly wired circuit will blow the fuse!

In Part 2, you will investigate a varactor diode. A varactor acts as a small capacitor for applications such as tuners. The *n*- and *p*-layers of the diode are conductive regions; the depletion region, with its absence of charge, acts as the dielectric. The capacitance varies inversely with the width of the depletion region, which is controlled by reverse bias. You will construct a resonant circuit using a varactor and measure the resonant frequency as the reverse bias voltage is changed.

Part 3 introduces two related photo sensors. The photodiode is a light sensor with a transparent window that allows light to enter. The light strikes the depletion region, causing hole-electron pairs to be formed and produces current in the external circuit. The photodiode is operated with reverse-bias. The second sensor is a *phototransistor* that is essentially a photodiode coupled to an internal transistor. This increases the sensitivity to light but the phototransistor is slower than a photodiode alone. You will use the phototransistor as a sensor to test the radiation pattern of an LED.

Reading

Floyd, *Electronic Devices*, Eighth Edition, Chapter 3

Key Objectives

Part 1: Use an oscilloscope to plot the characteristic curve of a zener diode. Test a zener regulator circuit for the effect of a changing source and a changing load. Test an IC regulator in a small power supply.

Part 2: Measure the resonant frequency and *Q* of a voltage-controlled resonant circuit containing a varactor diode as a function of the bias voltage.

Part 3: Measure the *V-I* characteristic for three light-emitting diodes (LEDs) and for a photocell. Measure the light from an LED as a function of direction.

Components Needed

Part 1: The Zener Diode and Regulator

> Resistors: one 220 Ω, one 1.0 kΩ, one 2.2 kΩ
> Four rectifier diodes 1N4001
> Capacitors: one 0.01 µF, one 220 µF capacitor
> One 1.0 kΩ potentiometer
> One 5 V zener, (1N4733A or equivalent)
> One regulator 7805 or 78L05
> One 12.6 V ac center-tapped transformer with fused line cord

Part 2: The Varactor Diode

> One 5.1 MΩ resistor
> One 10 kΩ potentiometer
> One MV2109 varactor
> One 15 mH inductor
> One 0.1 µF capacitor

Part 3: Light-Emitting Diode and Photodiode

> Resistors: one 510 Ω, one 1.0 kΩ, one 330 kΩ, one 1.0 MΩ
> Three LEDs, one red, one yellow, one green
> Light source (bright lamp or flashlight)
> One MRD500 photodiode
> One MRD300 phototransistor
> One 12.6 V ac center-tapped transformer with fused line cord
> Masking tape
> Heat shrink tubing (for light baffle) that fits over phototransistor
> Soldering iron

Part 1: The Zener Diode and Regulator

Safety note: You are connecting a 12.6 VAC transformer into the ac utility voltage. Be certain the line cord is insulated and fused. At no time should you touch the circuit when it is connected. You will make no connections or measurements on the primary side of the transformer.

1. Measure and record the values of the resistors listed in Table 3-1.

    Table 3-1

Resistor	Listed Value	Measured Value
R_1	220 Ω	
R_2	1.0 kΩ	
R_L	2.2 kΩ	

2. Observe the zener characteristic curve by setting up the circuit shown in Figure 3-1. Put scope in the X-Y mode. Sketch the *V-I* curve in Plot 3-1. The 1.0 kΩ resistor changes the scope's *y*-axis into a current axis (1 mA per volt). Label your plot for current and voltage.

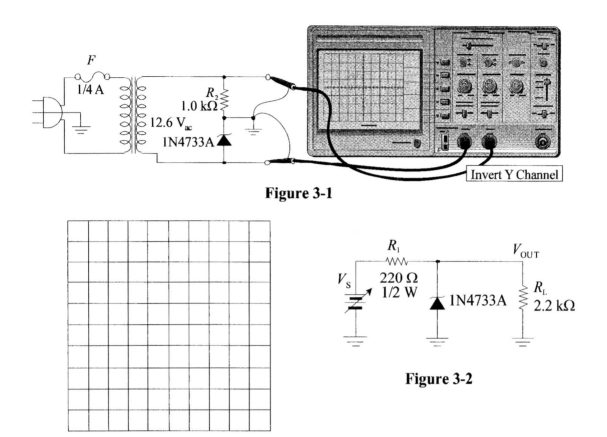

Figure 3-1

Plot 3-1

3. Investigate the effect of a zener regulator as the source voltage is varied. Connect the circuit shown in Figure 3-2. Set V_s to each voltage listed in Table 3-2 and measure the output (load) voltage, V_{out}.

Table 3-2

V_s	V_{out} (measured)	I_L (computed)	V_{R1} (computed)	I_s (computed)	I_Z (computed)
2.0 V					
4.0 V					
6.0 V					
8.0 V					
10.0 V					

4. From the measurements in step 3, complete Table 3-2. Apply Ohm's law to compute the load current, I_L, for each setting of the source voltage. The voltage across R_1 (V_{R1}) can be found by applying Kirchhoff's Voltage Law (KVL) to the outside loop. It is the difference between the source voltage, V_s, and the output voltage, V_{out}. Note that I_s is through R_1 and can be found using Ohm's law. Find

the zener current, I_Z, by applying Kirchhoff's Current Law (KCL) to the junction at the top of the zener diode.

What happens to the zener current after the breakdown voltage is reached?

---

5. Test the effect of a variable load resistance on a zener regulator. Construct the circuit shown in Figure 3-3. Set the power supply to a fixed +12.0 V output and adjust the potentiometer (R_L) for maximum resistance.

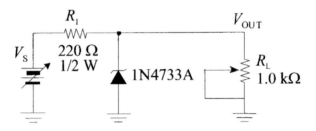

Figure 3-3

6. With the potentiometer set to 1.0 kΩ (maximum resistance), measure the load voltage (V_{out}) and record the voltage in Table 3-3. Compute the other parameters listed on the first row as before. (Use Ohm's law for I_L, KVL for V_{R1}, Ohm's law for I_s, and KCL for I_Z).

Table 3-3

R_L	V_{out} (measured)	I_L (computed)	V_{R1} (computed)	I_s (computed)	I_Z (computed)
1.0 kΩ					
750 Ω					
500 Ω					
250 Ω					
100 Ω					

7. Set the potentiometer to each value listed in Table 3-3 and repeat step 6.

8. From the data in Table 3-3, plot the output voltage as a function of load resistance in Plot 3-2. Choose a reasonable scale factor for each axis and add labels and a title to the plot.

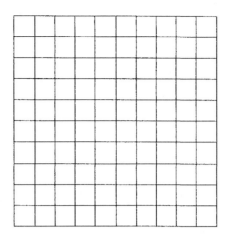

Plot 3-2

9.   Based on your results, what is the smallest load resistor that can be used and still maintain regulation?

_____

10. The pin outs for the 7805 and the 78L05 regulators are shown in Figure 3-4; either one of these can be used in this experiment. A bridge rectifier circuit with an IC regulator is shown in Figure 3-5. The circuit is similar to the bridge circuit in Experiment 2, Part 1, but with the addition of the three-terminal IC regulator, a larger filter capacitor and a small capacitor across the output. The transformer has a 12.6 V_{rms} secondary. Notice that <u>no</u> terminal on the transformer is grounded because of the bridge configuration. The center tap (if present) is not connected.

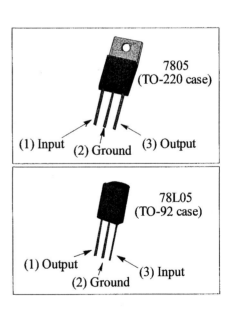

Figure 3-4

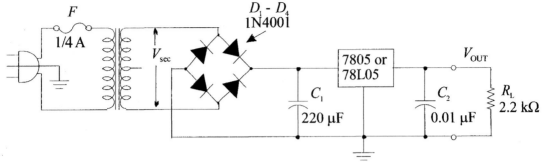

Figure 3-5

11. The input to the regulator is first filtered by the 220 µF filter capacitor. The 0.01 µF capacitor provides a path for noise to ground to reduce any "spikes" in the output. Connect the circuit, and measure the output ripple. If you did the bridge circuit in Experiment 2, compare your result with the unregulated bridge circuit you tested in step 9 of that experiment (note differences in capacitors).

Observations:

_____

_____

Conclusion: Part 1

Questions: Part 1

1. Observe the characteristic curve for a zener in Plot 3-1.
    (a) What portion of the curve is approximated by an open circuit?

    (b) What portion of the curve is approximated by a short circuit?

2. Line regulation of a zener regulator is normally expressed as a percentage and is given by the equation:

$$\text{Line regulation} = \frac{\Delta V_{OUT}}{\Delta V_{IN}} \times 100\%$$

Compute the line regulation expressed as a percentage for the circuit in Figure 3-2 using the data for the *last* two rows of Table 3-2. (Note that V_{IN} in the equation is equivalent to V_S in the table.)

3. Load regulation of a zener regulator, expressed as a percentage, is given by the equation:

$$\text{Load regulation} = \frac{V_{NL} - V_{FL}}{V_{FL}} \times 100\%$$

Compute the load regulation for the circuit in Figure 3-3. (Assume V_{OUT} for the 1.0 kΩ potentiometer at maximum setting = V_{NL} and V_{OUT} for the potentiometer at the 100 Ω setting represents V_{FL}).

Part 2: The Varactor Diode

1. Construct the varactor resonant circuit shown in Figure 3-6. The circuit uses a reverse-biased varactor to control the resonant frequency. C_1 isolates the bias voltage from ground. Adjust the bias voltage to zero volts. Using the oscilloscope, set the function generator to 1.0 V_{pp} at a frequency of 100 kHz. Observe the signal across the inductor. At this frequency, the amplitude will be quite small. Increase the frequency slowly, and you should observe the signal rise suddenly to a maximum, then fall. The frequency of the maximum signal is the resonant frequency. Measure the resonant frequency and record it in the first space in Table 3-4.

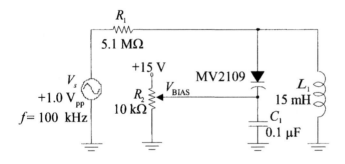

Figure 3-6

Table 3-4

V_{BIAS}	Resonant Frequency, f_r
0.0 V	
1.0 V	
2.0 V	
4.0 V	
8.0 V	
15.0 V	

2. Increase the bias voltage to +1.0 V. Measure and record the new resonant frequency. Continue in this manner for each bias voltage given in Table 3-4.

3. Plot the response of the circuit in Plot 3-3. The independent variable is bias voltage so it is plotted on the x-axis as shown. The resonant frequency is plotted on the y-axis. Add the proper labels and title to your graph.

25

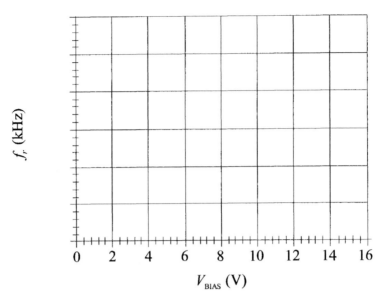

Plot 3-3

4. Measure the Q of the circuit. To do this, leave the bias voltage at a constant +15 V. Record the resonant frequency, f_r, you found previously for +15 V of bias in Table 3-5. Then raise the frequency of the generator until the voltage across the inductor drops to 70.7% of the maximum (at resonance). This is the upper critical frequency, f_{cu}. Measure and record this frequency in Table 3-5.

Table 3-5

Parameter	Measured Value
Resonant frequency, f_r	
Upper critical frequency, f_{cu}	
Lower critical frequency, f_{cl}	
Bandwidth, BW	
Q	

5. Reduce the frequency of the generator until the voltage across the inductor drops to 70.7% of the maximum. This is the lower critical frequency, f_{cl}. Measure and record this frequency in Table 3-5. Then compute the bandwidth, BW, for the circuit by subtracting f_{cl} from f_{cu}. Find Q for the circuit from the relation:

$$Q = \frac{f_r}{BW}$$

Record the calculated BW and Q in Table 3-5.

Conclusion: Part 2

Questions: Part 2
1. When the bias is 4 V, what is the total capacitance of the resonant circuit?

2. What evidence did you see in the experiment that the capacitance of the varactor decreases with increasing bias?

3. Assume that resistor R_1 in Figure 3-6 was changed to 100 kΩ.
   (a) Would this have an effect on the resonant frequency? Explain.

   (b) Would this have an effect on Q? Explain.

Part 3: Light-Emitting Diode and Photodiode

1. Test the *V-I* characteristic of three different colored LEDs: red, yellow, and green. Do this by setting up the circuit shown in Figure 3-7. Note that the transformer secondary is *not* grounded. With the power off to the circuit, place the oscilloscope in the X-Y mode with Channel 2 inverted. When a dot is on the screen, you need to keep the intensity *very low* to avoid damage to the CRT. Position the dot at center screen. Set the X channel to 1.0 V/DIV and set the Y channel to 5.0 V/DIV. The X channel represents voltage across the diode and the Y channel represents current. Observe the *V-I* characteristic for each of the three LEDs and describe your observations. Note the variation in the forward-bias voltage and the slope of the forward-bias characteristic.

Multisim

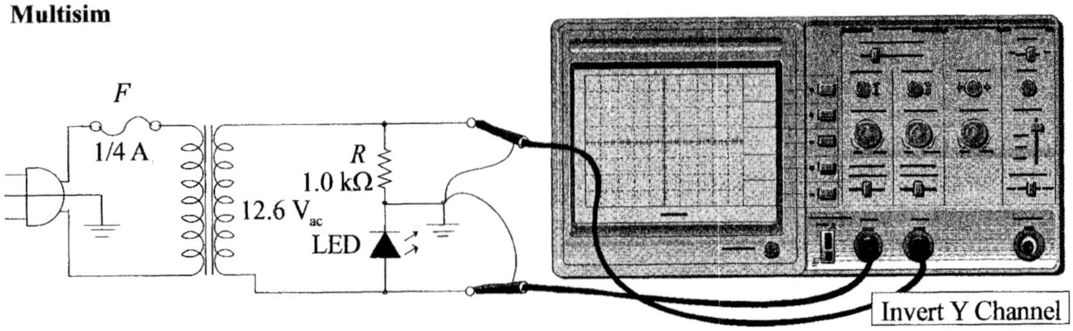

F

1/4 A

12.6 V$_{ac}$

R
1.0 kΩ

LED

Invert Y Channel

Figure 3-7

Red LED:

_____

Yellow LED:

_____

Green LED:

_____

2. You can observe the characteristics of a photodiode with a circuit similar to the one you built in step 1. Since a typical photodiode sources current in the microamp region, change the resistor in Figure 3-7 to 1.0 MΩ and replace the LED with a MRD500 photodiode. (A larger resistor will give a higher voltage). Observe the reverse part of the characteristic curve as you shine a bright light into the photodiode. The photodiode is directional, so the light needs to be directly into the photodiode.

Observations:

_____

_____

3. In this step, you will test the directional characteristics of an LED by making a detector using a phototransistor. (A phototransistor is used because it is much more sensitive than a photodiode.) Solder two wires approximately 15 cm long to the collector and emitter of a phototransistor (the outside leads). Then place a 3 cm piece of heat shrink tubing (approximately 5 mm diameter) over the phototransistor to serve as a light collimator; light should reach the phototransistor only through the tubing as shown in Figure 3-8(a). You may want to *slightly* heat the shrink tube to the transistor.

The phototransistor detector circuit is shown in Figure 3-8(b). Notice that the base lead (center) is left open. The two leads from the phototransistor are brought to the protoboard to complete the circuit. Connect a voltmeter across the collector resistor, R_C.

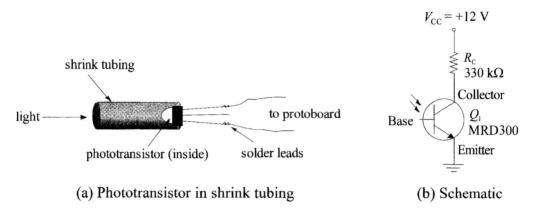

(a) Phototransistor in shrink tubing   (b) Schematic

Figure 3-8

4. Next, solder two wires approximately 15 cm long to the green LED. Then connect the circuit shown in Figure 3-9. The wires allow the LED to be positioned away from the circuit board. Figure 3-10 is marked for this experiment. Position the lighted LED over the spot marked for it. Tape it down with masking tape. Place the phototransistor on the 0° line aimed at the LED. You will need to have room lights dimmed. Record the voltage across the collector resistor. This voltage is proportional to the light from the LED.

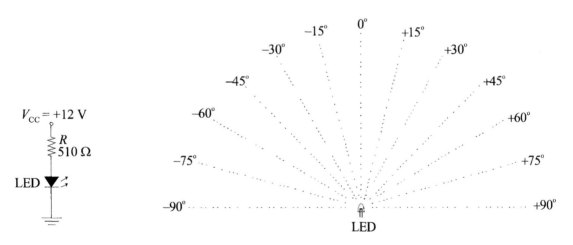

Figure 3-9   **Figure 3-10**

Keeping the phototransistor at a constant distance from the LED, measure the voltage across the collector resistor, R_C, every 15° around the LED. Set up a data table and record the values. Summarize your findings.

Conclusion: Part 3

Questions: Part 3

1. Consider the circuit shown in Figure 3-11. Assume the LED drops 2.0 V when it is forward-biased.

    (a) Compute the current in the LED.

    (b) Assume the maximum permissible current in the LED is 30 mA. What value of R will produce this current?

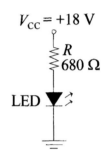

$V_{cc} = +18$ V

R
680 Ω

LED

Figure 3-11

2. LEDs are rated in terms of electrical and optical characteristics. Give examples of each characteristic that could be found on a specification sheet.

3. The photodiode characteristic was observed using a 1.0 MΩ resistor instead of the 1.0 kΩ resistor that was used for the LED. Explain why this change was necessary.

Multisim Simulation

Multisim

Multisim files for the lab manual are on the website www.prenhall.com/floyd. Open the Multisim file Experiment_03 IV Curve-nf and set up the Tektronix oscilloscope to view the IV curve for the LED. You will need to put the scope in the XY mode and invert the Y channel to see the curve in the correct orientation. Compare the simulation result with your experimental results. Replace the LED with a signal diode and compare the difference between it and an LED.

Experiment 4 Bipolar Junction Transistors

A bipolar junction transistor (BJT) is a three-terminal device capable of amplifying an ac signal. BJTs are current amplifiers. A small base current is amplified to a larger current in the collector-emitter circuit. An important BJT characteristic is the dc current gain, called the dc beta (β_{DC}), which is the ratio of collector current to base current. Another parameter is the ac beta (also called small signal current gain), which is a *change* in collector current divided by a *change* in base current. In Part 1 of this experiment, you will plot the characteristic curve for a BJT and use the plot to determine both the β_{DC} and the β_{ac}. You will also learn how to plot the curves automatically on an oscilloscope.

In Part 2, *switching circuits* used for control applications are introduced. Most digital applications use digital integrated circuits (ICs), but discrete transistor switching circuits are used when it is necessary to supply higher current or different voltages than can be furnished directly from the digital IC. Transistor switches are more reliable than mechanical switches, cost less, provide faster switching times, and can provide isolation when a load is in a dangerous or remote location.

In switching applications, transistors are normally operated in either cutoff or saturation. *Cutoff* refers to the condition where the transistor acts as an open switch; *saturation* occurs when the transistor acts as a closed switch. If a transistor is in cutoff, there is no current in the collector circuit; if it is saturated, it is conducting as much as possible.

After testing a basic one-transistor switching circuit, you will make three improvements: (1) avoiding the gradual switching by adding a second transistor, (2) raising the voltage threshold where switching occurs, and (3) adding hysteresis. Hysteresis refers to two different thresholds, depending on whether the output is already saturated or already in cutoff. The advantage of hysteresis is that the switching is less susceptible to noise.

Reading

Floyd, *Electronic Devices*, Eighth Edition, Chapter 4

Key Objectives

Part 1: Measure and graph the collector characteristic curves for a bipolar junction transistor; use this data to determine the β_{DC} of the transistor.

Part 2: Construct and test transistor switching circuits for its switching characteristics including the thresholds. Test a switching circuit with hysteresis.

Components Needed

Part 1: The BJT Characteristic Curve

Resistors: one 100 Ω resistor, one 33 kΩ resistor
One 2N3904 *npn* transistor (or equivalent)
One rectifier diode (1N4001 or equivalent)
One 12.6 V ac center-tapped transformer with fused line cord

Part 2: BJT Switching Circuits

Resistors: one 330 Ω, one 1.0 kΩ, two 10 kΩ
One 10 kΩ potentiometer
Two small signal *npn* transistors (2N3904 or equivalent)
One LED

Part 1: The BJT Characteristic Curve

1.   Measure and record the resistance of the resistors listed in Table 4-1.

Table 4-1

Resistor	Listed Value	Measured Value
R_1	33 kΩ	
R_2	100 Ω	

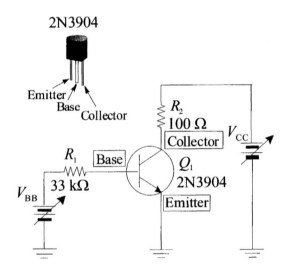

Figure 4-1

2.   Connect the common-emitter configuration illustrated in Figure 4-1. Start with both power supplies set to 0 V. The purpose of R_1 is to limit base current to a safe level and to allow indirect determination of the base current. Slowly increase V_{BB} until V_{R1} is 1.65 V. This sets up a base current of 50 µA, which can be shown by applying Ohm's law to R_1.

3.   Without disturbing the setting of V_{BB}, slowly increase V_{CC} until +2.0 V is measured between the transistor's collector and emitter. This voltage is V_{CE}. Measure and record V_{R2} for this setting. Record V_{R2} in Table 4-2 in the column labeled <u>Base Current = 50 µA</u>.

Table 4-2

V_{CE} (measured)	Base Current = 50 µA		Base Current = 100 µA		Base Current = 150 µA	
	V_{R2} (measured)	I_C (computed)	V_{R2} (measured)	I_C (computed)	V_{R2} (measured)	I_C (computed)
2.0 V						
4.0 V						
6.0 V						
8.0 V						

4. Compute the collector current, I_C, by applying Ohm's law to R_2. Use the measured voltage, V_{R2}, and the measured resistance, R_2, to determine the current. Note that the current in R_2 is the same as I_C for the transistor. Enter the computed collector current in Table 4-2 in the column labeled Base Current = 50 µA.

5. Without disturbing the setting of V_{BB}, increase V_{CC} until 4.0 V is measured across the transistor's collector to emitter. Measure and record V_{R2} for this setting. Compute the collector current by applying Ohm's law as in step 4. Continue in this manner for each of the values of V_{CE} listed in Table 4-2.

6. Reset V_{CC} for 0 V and adjust V_{BB} until V_{R1} is 3.3 V. The base current is now 100 µA.

7. Without disturbing the setting of V_{BB}, slowly increase V_{CC} until V_{CE} is 2.0 V. Measure and record V_{R2} for this setting in Table 4-2 in the column labeled Base Current = 100 µA. Compute I_C for this setting by applying Ohm's law to R_2. Enter the computed collector current in Table 4-2.

8. Increase V_{CC} until V_{CE} is equal to 4.0 V. Measure and record V_{R2} for this setting. Compute I_C as before. Continue in this manner for each value of V_{CC} listed in Table 4-2.

9. Reset V_{CC} for 0 V and adjust V_{BB} until V_{R1} is 4.95 V. The base current is now 150 µA.

10. Complete Table 4-2 by repeating steps 7 and 8 for 150 µA of base current.

11. Plot three collector characteristic curves using the data tabulated in Table 4-2. The collector characteristic curve is a graph of V_{CE} versus I_C for a constant base current. Choose a scale for I_C that allows the largest current observed to fit on the graph. Label each curve with the base current it represents. Graph the data on Plot 4-1 on the next page.

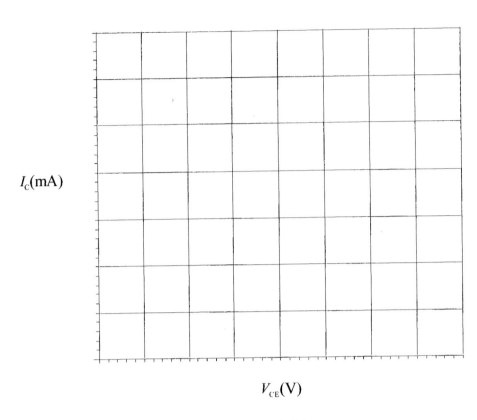

I_c(mA)

V_{CE}(V)

Plot 4-1

12. Use the characteristic curve you plotted to determine the ß$_{DC}$ for the transistor at a V_{CE} of 3.0 V and a base current of 50 μA, 100 μA, and 150 μA. Then repeat the procedure for a ß$_{DC}$ at a V_{CE} of 5.0 V. Tabulate your results in Table 4-3.

Table 4-3

	Current Gain, ß$_{DC}$		
V_{CE}	$I_B = 50$ μA	$I_B = 100$ μA	$I_B = 150$ μA
3.0 V			
5.0 V			

13. You can observe the collector curves, one at a time, on an oscilloscope. The circuit is a modification of the one used in the experiment and is shown in Figure 4-2. The collector supply is replaced with a low voltage transformer and diode. Start with V_{BB} set to 1.65 V as before ($I_B = 50$ μA).

Put the oscilloscope in X-Y mode and put both channels to the GND (ground) position. Keep the intensity low and position the dot in the lower left corner of the screen. Adjust the oscilloscope Y channel to 0.1 V/div (equivalent to 1.0 mA/div) and the X channel to 1 V/div and invert the Y channel.[1] Couple the signal to the scope and raise the intensity; you should see the first collector curve that you measured in the experiment. You can adjust V_{BB} to observe the other curves.

---

[1] If you cannot invert the Y channel, position the trace at the top of the screen. Increasing current will be toward the bottom of the screen.

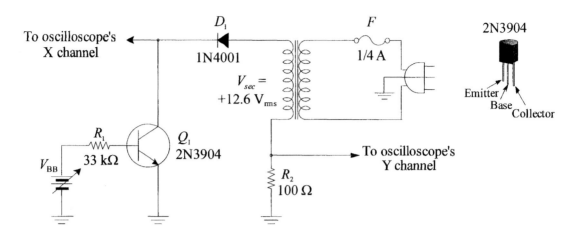

Figure 4-2

Conclusion: Part 1

Questions: Part 1

1. Does the experimental data indicate that β_{DC} is a constant at all points? Does this have any effect on the linearity of the transistor?

2. What effect would a higher β_{DC} have on the characteristic curves you measured?

3. What is the maximum power dissipated in the transistor for the data taken in the experiment?

4. (a) The dc alpha of a bipolar transistor is the collector current, I_C, divided by the emitter current, I_E. Using this definition and $I_E = I_C + I_B$, show that dc alpha can be written as:

$$\alpha_{DC} = \frac{\beta_{DC}}{\beta_{DC} + 1}$$

(b) Compute dc alpha for your transistor at $V_{CE} = 4.0$ V and $I_B = 100\ \mu$A.

5. What value of V_{CE} would you expect if the base terminal of a transistor were open? Explain your answer.

Part 2: BJT Switching Circuits

1. Measure and record the values of the resistors listed in Table 4-4. R_1 is a 10 kΩ potentiometer and is not listed in the table. R_{C1} is used in step 4.

Table 4-4

Resistor	Listed Value	Measured Value
R_B	10 kΩ	
R_C	1.0 kΩ	
R_{C1}	10 kΩ	
R_E	330 Ω	

2. Ideally, a transistor switching circuit should operate either in cutoff or in saturation. The circuit shown in Figure 4-3 is a basic transistor amplifier. It can easily be set for operation at these extremes by varying the potentiometer (R_1). However, it can also operate *between* cutoff and saturation, a condition not desirable in a switching circuit. Compute the collector-emitter voltage (V_{CE}) at cutoff and saturation and the voltage across the collector resistor at saturation (V_{RC}). Compute I_{sat} by assuming a 2.0 V drop across the LED and 0.1 V across the transistor. Show the computed values in Table 4-5.

36

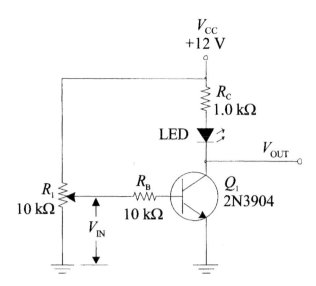

Table 4-5

Quantity	Computed Value	Measured Value
$V_{CE(cutoff)}$		
$V_{CE(sat)}$		
$V_{RC(sat)}$		
I_{sat}		

Figure 4-3

3. Construct the circuit in Figure 4-3 and observe the effect of varying the potentiometer. Set the potentiometer to the minimum value and measure V_{CE} at cutoff (the LED should be off). Then set the potentiometer for maximum (LED on), and measure V_{CE} (saturation) and the voltage across the 1.0 kΩ collector resistor, V_{RC} (saturation). The transistor is saturated since it can supply no more current in the collector circuit. Record measurements in Table 4-5 and compare to the computed values from step 2.

4. In this step, you will add a second transistor in a circuit that is similar to the Application Activity in the text. The additional gain due to a second transistor causes the switching action to improve dramatically. The circuit is shown in Figure 4-4. Notice that the 1.0 kΩ resistor is now the collector resistor for Q_2. The circuit works as follows. When V_{IN} is very low, Q_1 is off since it does not have sufficient base current. Q_2 will be in saturation because it can obtain ample base current through R_{C1}, so the LED is on. As the base voltage for Q_1 is increased, Q_1 begins to conduct. As Q_1 approaches saturation, the base voltage of Q_2 drops, causing it to go from a saturated to cutoff condition rapidly. The output voltage of Q_2 drops and the LED goes out. Construct the circuit, and test it as described in the next step.

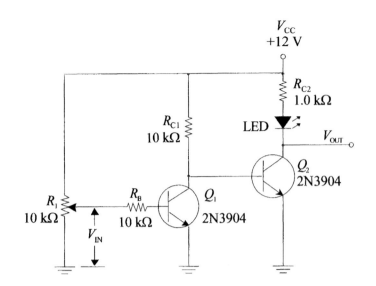

Table 4-6

Quantity	Measured Value
V_{IN} (LED on)	
V_{OUT} (LED on)	
V_{IN} (threshold)	
V_{OUT} (threshold)	

Figure 4-4

5. Set the potentiometer so that V_{IN} is a minimum (0 V). Since Q_1 is *off*, the LED should be *on*. Measure V_{IN} and V_{OUT} and record the readings in Table 4-6 in the first two rows. Monitor V_{IN} and *slowly* increase V_{IN} while watching the LED. You should observe that there is no dim condition for the LED – it will suddenly go off as the input voltage is increased. Record V_{IN} and V_{OUT} at the threshold where the LED just turns off. Notice that the output voltage indicates Q_2 is either in saturation or in cutoff.

6. In step 5, the switching threshold was distinct; however the threshold voltage is rather low and it is susceptible to noise on the input. There is another simple improvement you can make to this circuit. The improvement is shown in Figure 4-5. The common-emitter resistor, R_E, will raise the threshold voltage. In addition, because of the different saturation currents of the two transistors, the threshold will be different when the output is in cutoff than when the output is saturated. This is a very useful feature, called hysteresis, and is characteristic of Schmitt trigger circuits. Modify the circuit from step 5 by adding the 330 Ω resistor; then test it according to the procedure in the next step.

7. Set the potentiometer so that V_{IN} is a minimum (0 V). The LED should be *on*. Measure V_{IN} and V_{OUT} and record the readings in Table 4-7. The reading of V_{OUT} is higher than in the previous circuit but the transistor is still saturated (why?). Test the upper threshold by monitoring V_{IN} as you *slowly* increase V_{IN}. You will see the LED go out suddenly. Record this as the upper threshold voltage. Measure and record V_{OUT} and confirm that the transistor is in cutoff. Now *slowly* reduce V_{IN}. Notice that the LED stays out until the voltage is much lower. When the LED comes on, record V_{IN} and V_{OUT} at the lower threshold.

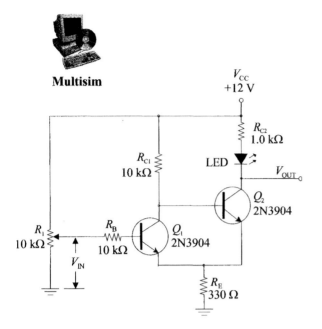

Multisim

V_{CC} +12 V

R_{C2} 1.0 kΩ

R_{C1} 10 kΩ

LED

V_{OUT}

Q_2 2N3904

R_1 10 kΩ

R_B 10 kΩ

Q_1 2N3904

V_{IN}

R_E 330 Ω

Table 4-7

Quantity	Measured Value
V_{IN} (LED on)	
V_{OUT} (LED on)	
V_{IN} (upper threshold)	
V_{OUT} (upper threshold)	
V_{IN} (lower threshold)	
V_{OUT} (lower threshold)	

Figure 4-5

8. A transfer curve is a plot of the output of a circuit (plotted along the y-axis) versus the input (plotted along the x-axis). Set up a data table for the circuit in Figure 4-5 and record the input and output voltage as the input is increased and decreased. Take sufficient data that you know precisely what the output voltage is for the range of input voltages. Then plot the transfer curve for the circuit in Plot 4-2. Label the axes of your plot.

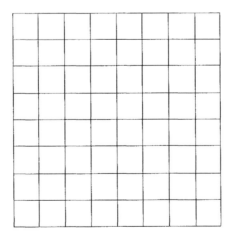

Plot 4-2

39

Conclusion: Part 2

Questions: Part 2

1. Give at least three advantages of transistor switching circuits.

2. What is the purpose of R_B in Figure 4-5?

3. Assume you wanted to determine the base current in Figure 4-5. What voltage measurement would you make to do this indirectly?

4. In step 6, it was stated that the saturation current for the two transistors is different. Why?

Multisim Simulation

Multisim

Multisim files for the lab manual are on the website www.prenhall.com/floyd. Open the Multisim file Experiment_04 Switch-nf. The input is set up a little differently than in Figure 4-5. Instead of controlling the input with a potentiometer, the input is a triangle wave that goes from 0-4 V. The Tektronix oscilloscope is used to monitor the input and output simultaneously. You will need to adjust the scope controls. Read the input thresholds by noting where the output changes with respect to the input triangle waveform. Compare the simulated result with your lab result.

Experiment 5 Transistor Bias Circuits

For a transistor to operate in the linear region, it is necessary to forward-bias the base-emitter junction and to reverse-bias the base-collector junction. The purpose of bias is to provide dc voltages to set up the quiescent conditions which establish the operating point on the characteristic curve for the device.

In Part 1, you will investigate three common bias circuits. The circuits are: (1) base bias and a form of base bias called emitter-feedback bias (2) voltage-divider bias, and (3) collector-feedback bias. You will test each of the three transistors, one at a time, in various bias circuits, starting with fixed base bias. These basic bias circuits are described in the text for *npn* transistors, which are used here. Other types of bias are sometimes used, but in normal linear operation, the key is that the bias network must forward-bias the base-emitter junction and reverse-bias the base-collector junction.

In Part 2, you will investigate emitter bias and voltage-divider bias with two supplies. Emitter bias is very stable but requires two power supplies – one positive and the other negative. A small change to the emitter bias circuit forms a voltage-divider bias with two supplies, which is also stable. The specific formulas for calculating this circuit are not given, but can easily be determined by applying basic rules that you are familiar with. This circuit can then be converted to a sensitive circuit for monitoring temperature.

Reading
Floyd, *Electronic Devices*, Eighth Edition, Chapter 5

Key Objectives
Part 1: Construct and analyze three types of transistor bias circuits: base bias, voltage-divider bias, and collector-feedback bias.

Part 2: Construct and analyze emitter bias, voltage-divider bias with two supplies, and a thermistor circuit.

Components Needed
Part 1: Three Bias Circuits
Resistors (one of each): 680 Ω, 1.5 kΩ, 2.0 kΩ, 6.8 kΩ, 33 kΩ, 360 kΩ, 1.0 MΩ

Three small signal *npn* transistors, (2N3904 or equivalent)

Part 2: Emitter Bias and Two-Supply Voltage-Divider Bias
Resistors: one 330 Ω, one 470 Ω, two 4.7 kΩ

Potentiometer: 10 kΩ

Three small signal *npn* transistors, (2N3904 or equivalent)

One NTC 5 kΩ thermistor (Mouser part number 527-2006)

Part 1: Three Bias Circuits
Fixed Base Bias
1. Measure and record the values of the resistors listed in Table 5-1.

Table 5-1

Resistor	Listed Value	Measured Value
R_B	1.0 MΩ	
R_C	2.0 kΩ	
R_E	1.5 kΩ	

2. Fixed base bias will be investigated first, but it is the most sensitive to differences in β_{DC}. Transistors of the same type can have widely different values of β_{DC}, which generally restricts the fixed base bias to more specialized applications such as switching circuits such as you used in the last experiment. In step 4, you will modify the fixed base bias with an emitter resistor, which will add stability to this type of circuit. (R_E is not used in the first circuit.)

    The manufacturer's specification sheet for the 2N3904 shows β_{DC} can range from 100 to 300, a factor of 3, which implies that the collector current can also vary by a factor of 3 with base bias! Assuming the β_{DC} for a 2N3904 is in the middle of its specified range (200), compute the parameters listed in Table 5-2 for the fixed base bias circuit shown in Figure 5-1. Start by computing the voltage across the base resistor, V_{RB}, and the current in this resistor, I_B. Using β_{DC}, find the collector current, I_C, the voltage across the collector resistor, V_{RC}, and the voltage from collector to ground, V_C.

Table 5-2

DC Parameter	Computed Value	Measured Value		
		Q_1	Q_2	Q_3
V_{RB}				
I_B				
I_C				
V_{RC}				
V_C				

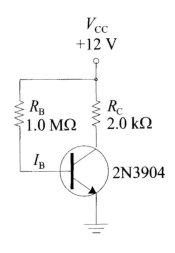

Figure 5-1

3. Label each of three *npn* transistors as Q_1, Q_2, and Q_3 in a manner that allows you to keep track of each transistor. Construct the circuit shown in Figure 5-1 using Q_1. Measure the voltages listed in Table 5-2 for Q_1. Then remove Q_1 from the circuit and test the other two transistors in the same circuit. Record the data in Table 5-2.

Emitter-Feedback Bias

4. Emitter-feedback bias is a form of base bias but with increased stability due to the addition of an emitter resistor. Add the emitter resistor to the base bias circuit and observe the results. Calculate the parameters in Table 5-3 for the circuit in Figure 5-2 and enter the calculations in the table. Use the method described in the text. Use β_{DC} as 200 for the calculated value.

5. Construct the circuit and measure the parameters in Table 5-3 for each of the three transistors you tested before.

Table 5-3

DC Parameter	Computed Value	Measured Value		
		Q_1	Q_2	Q_3
V_{RB}				
I_B				
I_C				
V_{RC}				
V_C			,	

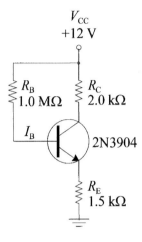

Figure 5-2

Voltage-Divider Bias

6. Although the emitter-stabilized base bias you tested in step 5 has better stability than fixed base bias, you may have noticed that parameters still vary between transistors. A more stable form of bias is voltage-divider bias, which you will test next using the same three transistors. The circuit meets the condition that $\beta R_E \geq 10 R_2$, so the approximate analysis method, given in Section 5-2 of the text, can be used.

Measure and record the values of the resistors listed in Table 5-4.

Table 5-4

Resistor	Listed Value	Measured Value
R_1	33 kΩ	
R_2	6.8 kΩ	
R_E	680 Ω	
R_C	2.0 kΩ	

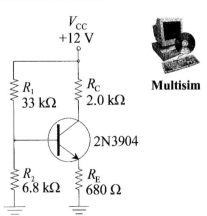

Multisim

Figure 5-3

7. Compute the parameters listed in Table 5-5 for the circuit shown in Figure 5-3 using the approximate method for all calculations.

Table 5-5

DC Parameter	Computed Value	Measured Value		
		Q_1	Q_2	Q_3
V_B				
V_E				
$I_E \approx I_C$				
V_{RC}				
V_C				

8. Substitute the same three transistors into the voltage-divider circuit and measure the parameters listed in Table 5-5. Record your data in the table.

Collector-Feedback Bias

9. Measure and record the values of the resistors listed in Table 5-6.

Table 5-6

Resistor	Listed Value	Measured Value
R_B	360 kΩ	
R_C	2.0 kΩ	

10. Compute the parameters listed in Table 5-7 for the circuit shown in Figure 5-4. The equations for collector-feedback bias are developed in the text on page 235. Notice there is no emitter resistor. For this case, the collector current is found from Equation 5-11. Assume β_{DC} is 200 for the calculation. Calculate the voltage across the collector resistor, V_{RC}, and the collector voltage, V_C based on this assumption.

11. Construct the circuit shown in Figure 5-4 using transistor Q_1. Measure the voltages listed in Table 5-7 for Q_1. Then remove Q_1 from the circuit and test the other two transistors in the same circuit. Record all measurements in Table 5-7.

Table 5-7

DC Parameter	Computed Value	Measured Value		
		Q_1	Q_2	Q_3
I_C				
V_{RC}				
V_C				

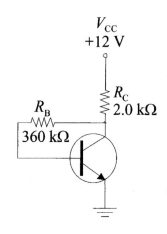

Figure 5-4

44

Conclusion: Part 1

Questions: Part 1

1. Compare your observations of the three bias methods tested in the experiment. Which showed the *least* variation among the transistors?

2. Would the addition of an emitter resistor to the circuit in Figure 5-4 improve the bias stability? Justify your answer.

Part 2: Emitter Bias and Two-Supply Voltage-Divider Bias

1. Measure and record the values of the resistors listed in Table 5-8. R_2 is not used in the first circuit.

Table 5-8

Resistor	Listed Value	Measured Value
R_B	4.7 kΩ	
R_C	330 Ω	
R_E	470 Ω	
R_2	4.7 kΩ	

2. Emitter bias is an excellent way of obtaining stable bias. It requires both a positive and negative power supply. Figure 5-5 shows a transistor with emitter bias. Construct the circuit and compute the two dc parameters given in Table 5-4. The simplest way to obtain an approximation of I_E is to assume the emitter is at a potential of −1 V and apply Ohm's law to R_E (this is valid if R_B is about 10X larger than R_E.) This approximation has been entered on Table 5-9 as the calculated value for the emitter voltage. (You can obtain a more exact value by applying Equation 5-7 in the text to the circuit). Enter the computed values in Table 5-9.

3. Measure V_E and V_C for the first transistor and enter the measured values in Table 5-9.

4. Replace the first transistor with the second transistor and repeat step 3.

5. Replace the second transistor with the third transistor and repeat step 3.

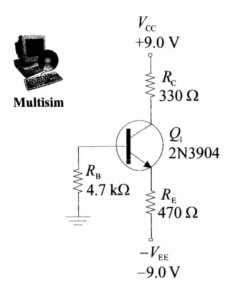

Multisim

Table 5-9

Quantity	Computed Value	Measured Value		
		Q_1	Q_2	Q_3
V_E	−1 V			
$I_E \approx I_C$				
V_C				

Figure 5-5

6. In this step, the circuit is changed to a combination of voltage-divider and emitter bias shown in Figure 5-6. The circuit is a minor modification of the previous circuit, producing very stable bias. Notice that R_B is renumbered as R_2 and connected to $-V_{EE}$. Calculate the base voltage; then subtract V_{BE} to obtain V_E. Calculate I_E and V_C, assuming that $I_E \approx I_C$. Measure the dc voltages that are listed in Table 5-10 with all three transistors.

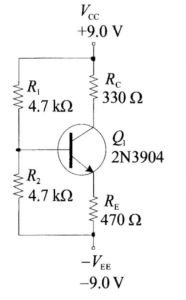

Table 5-10

Quantity	Computed Value	Measured Value		
		Q_1	Q_2	Q_3
V_E				
$I_E \approx I_C$				
V_C				

Figure 5-6

7. In this step, you will modify the circuit to act as a temperature sensor. This circuit is similar, but not identical to the Application Assignment in the text. To simplify the application, this circuit indicates temperature with an LED instead of a voltage level and has a potentiometer to enable you to set it for different ambient conditions or sensors. The circuit has also been modified from the one in the text to be much more sensitive to a bias change by putting the emitter on ground and biasing the base near 0 V.

    To limit the current in the LED to a reasonable level, move the 470 Ω resistor from the emitter to the collector as shown in Figure 5-7 (now labeled as R_C). The 10 kΩ potentiometer is used to set the threshold level. Construct the circuit.

8. Adjust the potentiometer so that the LED is on, but near the threshold. Then grasp the thermistor with your fingers and hold it for a few seconds.

    Observations:

    _____

    _____

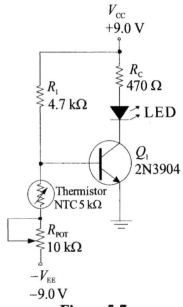

Figure 5-7

Conclusion: Part 2

Questions: Part 2
1. Give at least one advantage and one disadvantage to emitter bias.

47

2. When is it valid to assume the emitter voltage is –1 V in an emitter-biased circuit?

3. What change would you make to the circuit in Figure 5-7 to cause the LED to turn on with increasing temperature?

Multisim Simulation

Multisim

Multisim files for the lab manual are on the website www.prenhall.com/floyd. Open the Multisim file Experiment_05_BJT_Bias-pt1. There are two circuits that are the same as Figure 5-3 with faults. Using a dc voltmeter, determine the fault in each circuit.

Then open the Multisim file Experiment_05_BJT_Bias-pt2. There are two circuits that are the same as Figure 5-5 with faults. Using a dc voltmeter, determine the fault in each circuit.

Experiment 6 BJT Amplifiers

This experiment focuses on small-signal amplifiers that produce a replica of the input signal at the output. In Part 1, you will test a common-emitter amplifier and compare your results to the calculated and simulated values. In a common-emitter (CE) amplifier, the input signal is applied between the base and emitter, and output signal is developed between the collector and emitter. The transistor's *emitter* is common to the input and output circuits, hence, the term common-emitter.

In Part 2, you will investigate the common-collector (CC) amplifier using a *pnp* transistor. The common-collector (CC) amplifier (also called the *emitter-follower*) has the input signal applied to the base, and the output signal is taken from the emitter. The ac output voltage almost perfectly duplicates the input voltage waveform. While this implies that the voltage gain is approximately 1, the current gain is not; hence, it can deliver increased signal power to a load. The CC amplifier is characterized by a high input resistance and a low output resistance.

In part 3, two CE amplifiers are combined into a two stage preamplifier, similar to preamp in the Application Activity in the text. The input frequency is specified to be higher in the experiment because of an optional exercise, in which automatic gain control (AGC) is described. Automatic gain control means the circuit will set the actual gain depending on the level of the input signal. It is usually applied to the higher frequency circuits found in communication receivers. Another amplifier, the differential amplifier, is discussed in this chapter. This circuit is investigated in Chapter 12 in conjunction with operational amplifiers.

Reading
Floyd, *Electronic Devices*, Eighth Edition, Chapter 6

Key Objectives
Part 1: Compute the dc and ac parameters for a common-emitter (CE) amplifier. Build the amplifier and measure these parameters.
Part 2: Compute the dc and ac parameters for a common-collector (CC) amplifier using a *pnp* transistor. Build the amplifier and measure these parameters.
Part 3: Construct a two-stage transistor amplifier and measure the dc and ac parameters including the input resistance, output resistance, voltage gain, and power gain.

Components Needed
Part 1: The Common-Emitter Amplifier
Resistors: one 100 Ω, one 330 Ω, two 1.0 kΩ, one 4.7 kΩ, two 10 kΩ
Capacitors: two 1.0 µF, one 47 µF
One 10 kΩ potentiometer
One 2N3904 *npn* transistor

Part 2: The Common-Collector Amplifier

Resistors: two 1.0 kΩ, one 10 kΩ, one 33 kΩ
Capacitors: one 1.0 µF, one 10 µF
One 10 kΩ potentiometer
One 2N3906 *pnp* transistor

Part 3: Multistage Amplifiers

Resistors:
one of each: 220 Ω, 1.0 kΩ, 2.0 kΩ, 4.7 kΩ, 6.8 kΩ, 10 kΩ, 33 kΩ, 47 kΩ
two of each: 22 kΩ, 100 kΩ, 330 kΩ
One 2N3904 *npn* transistor (or equivalent)
One 2N3906 *pnp* transistor (or equivalent)
One 100 kΩ potentiometer
Capacitors: two 0.1 µF, three 1.0 µF
Optional AGC exercise:
One MPF102 *n*-channel JFET
One 1N914 signal diode (or equivalent)
Resistors: one 100 Ω, one 220 kΩ
Capacitors: one additional 1.0 µF capacitor

Part 1: The Common-Emitter Amplifier

1.   Measure and record the values of the resistors listed in Table 6-1.

Table 6-1

Resistor	Listed Value	Measured Value
R_1	10 kΩ	
R_2	4.7 kΩ	
R_{E1}	100 Ω	
R_{E2}	330 Ω	
R_C	1.0 kΩ	
R_L	10 kΩ	

Table 6-2

DC Quantity	Computed Value	Measured Value
V_B		
V_E		
I_E		
V_C		
V_{CE}		

2.   Calculate the dc quantities listed in Table 6-2 for the CE amplifier in Figure 6-1. Because $\beta R_E \geq 10R_2$, you can use the unloaded voltage-divider analysis method to obtain V_B. Subtract V_{BE} from V_B to obtain V_E. Compute I_E by applying Ohm's law to the resistors in the emitter circuit. Find V_C by subtracting V_{RC} from V_{CC}. Subtract V_E from V_C to obtain V_{CE}. Enter your computed values in Table 6-2.

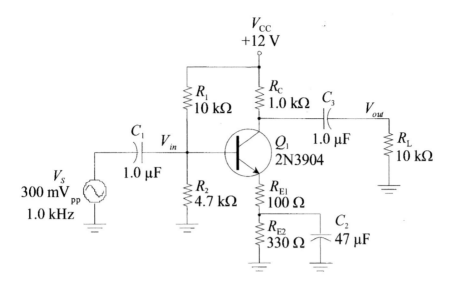

Figure 6-1

3. Construct the amplifier shown in Figure 6-1. The signal generator should be turned off. Measure and record the dc voltages listed in Table 6-2.

4. Calculate the four ac parameters listed in the first column of Table 6-3. The input signal, V_{in}, is given as 300 mV$_{pp}$. This is both V_{in} and the ac base voltage, V_b. Multiply V_{in} by the computed voltage gain to calculate the ac voltage at the collector; this is both V_c and V_{out}.

5. Turn on the signal generator and set V_{in} for 300 mV$_{pp}$ at 1.0 kHz with the generator connected to the circuit. Use the oscilloscope to set the proper voltage and check the frequency. Measure the ac signal voltage at the transistor's emitter and at the collector. Note that the signal at the emitter is less than at the base. Use V_{in} and the ac collector voltage (V_{out}) to determine the measured voltage gain, A_v. The measurement of $R_{in(tot)}$ and β_{ac} is explained in step 6 and 7. Record the measured values of V_{in}, V_e, V_{out}, and A_v in Table 6-3.

Table 6-3

AC Quantity	Computed Value	Measured Value
$V_{in} = V_b$	300 mV$_{pp}$	
V_e		
r_e		
$V_{out} = V_c$		
A_v		
$R_{in(tot)}$		
β_{ac}		

6. The measurement of $R_{in(tot)}$ is done indirectly because it is an ac resistance that cannot be measured with an ohmmeter. The output signal (V_{out}) is measured with an oscilloscope and recorded with the amplifier operating normally (no clipping or distortion). A rheostat (R_{test}) is then inserted in series with the source as shown in Figure 6-2. The rheostat is varied until V_{out} drops to one-half the value prior to inserting R_{test}. With this condition, $V_{in} = V_{test}$ and $R_{in(tot)}$ must be equal to R_{test}. R_{test} can then be removed and measured with an ohmmeter. Using this method, measure $R_{in(tot)}$ and record the result in Table 6-3.

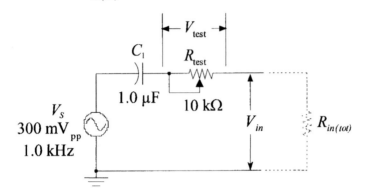

Figure 6-2 Measurement of $R_{in(tot)}$.

7. You can obtain an estimate of β_{ac} from $R_{in(tot)}$ and the known bias resistor values. Recall that $R_{in(tot)} = (\beta_{ac}(R_{E1} + r_e')) \| R_1 \| R_2$. Starting with the parallel resistor formula, you can rearrange it and use it to calculate β_{ac} indirectly. The equation is:

$$\beta_{ac} = \cfrac{1}{\cfrac{\left(R_{E1} + r_e'\right)}{R_{in(tot)}} - \cfrac{\left(R_{E1} + r_e'\right)}{R_1} - \cfrac{\left(R_{E1} + r_e'\right)}{R_2}}$$

Enter the result as the measured value of β_{ac} in Table 6-3. The result is an approximation of β_{ac}.

Troubleshooting

8. Remove the bypass capacitor, C_2, from the circuit, simulating an open capacitor. Measure the ac signal voltage at the transistor's base, emitter, and collector. Measure the voltage gain of the amplifier.
Observations with open C_2:

_____

9. Replace C_2 and reduce R_L to 1.0 kΩ, simulating a change in load conditions. Observe the ac signal voltage at the transistor's base, emitter, and collector.
Observations for smaller load:

_____

10. Replace R_L with the original 10 kΩ resistor and open R_{E1}. Measure the dc voltages at the base, emitter, and collector. Is the transistor in cutoff or in saturation? Explain.

_____

11. Replace R_{E1} and open R_2. Measure the dc voltages at the base, emitter and collector. Is the transistor in cutoff or saturation? Explain.

_____

Conclusion: Part 1

Questions: Part 1

1. In step 6, you were instructed to measure the input resistance while monitoring the output voltage. Why is this procedure better than monitoring the base voltage?

2. Does the load resistor have any effect on the input resistance? Explain your answer.

3. What is the purpose of the unbypassed emitter resistor R_{E1}? What design advantage does it offer?

4. When the bypass capacitor, C_2, is open, you found that the gain is affected. Explain why.

Part 2: The Common-Collector Amplifier

1. Test a CC amplifier (also called an emitter-follower) constructed with a *pnp* transistor. Measure and record the values of the resistors listed in Table 6-4.

Table 6-4

Resistor	Listed Value	Measured Value
R_1	33 kΩ	
R_2	10 kΩ	
R_E	1.0 kΩ	
R_L	1.0 kΩ	

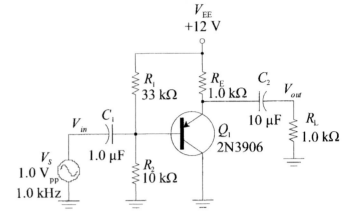

Figure 6-3

2. Compute the dc quantities listed in Table 6-5 for the CC amplifier shown in Figure 6-3. Remember that the emitter voltage is 0.7 V larger than V_B for the *pnp* transistor. Enter your computed dc quantities in Table 6-5.

Table 6-5

DC Quantity	Computed Value	Measured Value
V_B		
V_E		
I_E		
V_{CE}		

Table 6-6

AC Quantity	Computed Value	Measured Value
V_b	1.0 Vpp	
V_e		
r_e		
A_v		
$R_{in(tot)}$		
A_p		

3. Construct the amplifier shown in Figure 6-3. The signal generator should be turned off. With the power supply on, measure and record the dc voltages listed in Table 6-5. Your measured and computed values should agree within 10%.

4. Compute and record the ac quantities listed in Table 6-6. The emitter resistance is found using Equation 6-1 in the text. Assume V_b is the same as the source voltage, V_s. If you do not know the β_{ac} for your transistor, you can still obtain reasonable results if you assume it is 100.

5. Turn on the signal generator and set V_s for 1.0 Vpp at 1.0 kHz. Use the oscilloscope to set the proper voltage and check the frequency. Measure the input ac signal voltage at the base, V_b, and the output signal voltage at the emitter, V_e, to determine the voltage gain, A_v. Measure $R_{in(tot)}$ using the method employed for the

54

CE amplifier described in Part 1 (step 6). To calculate power, substitute the measured $R_{in(tot)}$ and measured R_L value into the power formula V^2/R (use rms voltages to calculate power). Record the measured ac parameters in Table 6-6.

6. With a two-channel oscilloscope, compare the input and output waveforms. What is the phase relationship between V_{in} and V_{out}?

Troubleshooting

7. Table 6-7 lists some possible troubles with the CC amplifier. For each trouble listed, predict the effect on the dc voltages. Then insert the trouble into the circuit and test your prediction. Insert the open collector and open emitter troubles by removing the transistor lead and measuring the voltages at the circuit. For each fault, describe the effect on the ac output waveform (clipped, no output, etc.).

Table 6-7

Trouble	DC Predictions			DC Measurements			Effect of Trouble on V_{out}
	V_B	V_E	V_{CE}	V_B	V_E	V_{CE}	
R_1 open							
R_2 open							
R_1 shorted							
R_E open							
open collector							
open emitter							

8. Replace R_L with a 10 kΩ variable resistor set to 1.0 kΩ. Connect an oscilloscope probe to the emitter. Increase the signal until you just begin to observe clipping. If the positive peaks are clipped, you are observing *cutoff* clipping because the transistor is turned off. If the negative peaks are clipped, this is called *saturation* clipping because the transistor is fully conducting. What type of clipping is first observed?

---

9. Vary R_L while observing the output waveform. Describe your observations.

Conclusion: Part 2

Questions: Part 2

1. In step 8, you observed the effect of clipping due to saturation or cutoff of the transistor. The statement was made that if the positive peaks are clipped, you are observing *cutoff* clipping because the transistor is turned off. Is this statement true if the CC circuit had been constructed with an *npn* transistor? Why or why not?

2. Common-collector amplifiers have a voltage gain less than 1 but still provide power gain. Explain why.

Application Activity

Part 3: Multistage Amplifiers

1. Measure and record the values of the resistors listed in Table 6-8.

Table 6-8

Resistor	Listed Value	Measured Value
R_A	100 kΩ	
R_B	2.0 kΩ	
R_1	330 kΩ	
R_2	330 kΩ	
R_{E1}	33 kΩ	
R_{E2}	1.0 kΩ	
R_{C1}	22 kΩ	
R_3	47 kΩ	
R_4	22 kΩ	
R_{E3}	4.7 kΩ	
R_{E4}	220 Ω	
R_{C2}	6.8 kΩ	
R_L	10 kΩ	

Table 6-9

DC Parameter	Computed Value	Measured Value
$V_{B(Q1)}$		
$V_{E(Q1)}$		
$I_{E(Q1)}$		
$V_{C(Q1)}$		
$V_{CE(Q1)}$		
$V_{B(Q2)}$		
$V_{E(Q2)}$		
$I_{E(Q2)}$		
$V_{C(Q2)}$		
$V_{CE(Q2)}$		

2. Figure 6-4 shows the two-stage amplifier for this part of the experiment that is
    similar to the text Application Activity. The input voltage divider is not part of the
    amplifier; it only serves to reduce the signal from your function generator by a
    known amount. Compute the dc parameters listed in Table 6-9. The steps are:

    (a) Mentally remove (open) capacitors from the circuit since they appear open
         to dc. Solve for the base voltage, V_B of Q_1. By inspection, the dc base
         voltage is zero; however, if the resistors are not equal, the base voltage can
         be found by applying the voltage-divider rule and the superposition theorem
         to R_1 and R_2.

    (b) Add the 0.7 V forward-bias drop across the base-emitter diode of Q_1 from
         V_B to obtain the emitter voltage, V_E, of Q_1.

    (c) Find the voltage across the emitter resistors and apply Ohm's law to solve
         for the emitter current, I_E, of Q_1.

    (d) Assume the emitter current (step 2c) is equal to the collector current, I_C, of
         Q_1. Find the voltage across R_{C1} using Ohm's law. V_C is found by
         algebraically adding this voltage to V_{EE}. Solve for V_{CE} of Q_1 by subtracting
         V_E from V_C (note that this is negative because of the *pnp* transistor).

    (e) Compute the base voltage of Q_2 by applying the superposition theorem and
         voltage divider rule to R_3 and R_4. Subtract 0.7 V from the base voltage of Q_2
         to find the emitter voltage of Q_2. Find the voltage across the emitter resistors
         and apply Ohm's law to determine the emitter current in Q_2.

    (f) Assume the emitter current (step 2e) is equal to the collector current, I_C, of
         Q_2. Find the voltage across R_{C2} and use it to calculate V_C and V_{CE} for Q_2.

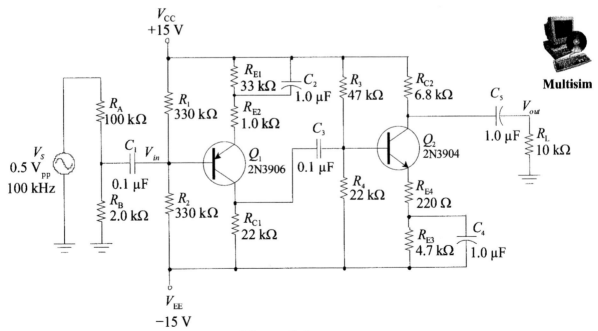

Figure 6-4

3. Construct the two stage amplifier. The function generator should be turned off. Measure and record the dc voltages listed in Table 6-9. Your measured and computed values should agree within 10%.

4. Calculate the ac parameters listed in Tables 6-10 and 6-11 using the lettered steps that follow. Table 6-10 only lists parameters that will not be measured later; Table 6-11 includes parameters that you will measure in steps 5, 6, and 7.

    (a) Mentally, replace all capacitors with a short. The ac resistance in the emitter circuit includes the unbypassed emitter resistor and the ac resistance of the transistor. Compute the ac emitter resistance of each transistor, r_e', from the equation:

    $$r_e' = \frac{25 \text{ mV}}{I_E}$$

    (b) Calculate the input and output resistance of Q_1. The input resistance is listed in Table 6-11. It includes the bias resistors in parallel with the ac resistance of the emitter circuit reflected into the base circuit. The output resistance is simply the value of the collector resistor.

    $$R_{in(Q1)} = R_1 \| R_2 \| \left\{ \beta_{ac} \left(r_e' + R_{E2} \right) \right\}$$

    (c) Calculate the input and output resistance of Q_2. As before, the input resistance includes the bias resistors and the ac emitter resistance reflected to the base circuit. The output resistance is again the collector resistor and is listed in Table 4-11.

    $$R_{in(Q2)} = R_3 \| R_4 \| \left\{ \beta_{ac} \left(r_e' + R_{E4} \right) \right\}$$

    (d) Calculate the unloaded gain, $A_{v(NL)}$, of each stage. The unloaded voltage gain for the common-emitter transistors can be written:

    $$A_{v(NL)} = \frac{V_{out}}{V_{in}} = \frac{I_c R_C}{I_e \left(r_e' + R_{e(ac)} \right)} = \frac{R_C}{\left(r_e' + R_{e(ac)} \right)}$$

    (e) Calculate the overall gain, A_v'. It is easier to calculate the voltage gain of a multistage amplifier by computing the *unloaded* voltage gain for each stage. The loading effect is found by computing a voltage-divider for the output resistance of one stage and the input resistance for the next stage (see Figure 6-5). Each transistor is drawn as an amplifier consisting of an input resistance, R_{in}, an output resistance, R_{out}, along with its unloaded gain, $A_{v(NL)}$. Then, the overall gain, A_v', of this amplifier (with no load resistor) can be found by:

    $$A_v' = A_{v(NL)(Q1)} \left(\frac{R_{in(Q2)}}{R_{out(Q1)} + R_{in(Q2)}} \right) A_{v(NL)(Q2)}$$

Enter the calculated overall gain in the first line of Table 6-11.

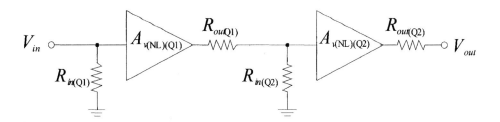

Figure 6-5

(f) Using the calculated overall gain, compute the expected output voltage for an input of 10 mV. Enter the calculated $V_{out(Q2)}$ on the last line of Table 6-11.

5. In this step, and the following two steps, you will measure parameters for the amplifier. Connect the function generator voltage to the divider composed of R_A and R_B as shown in Figure 6-4. (As mentioned, these resistors serve only to attenuate the generator signal by a known amount; they are not part of the amplifier.) Turn on the function generator and set V_s for a 0.5 V_{pp} sine wave at 100 kHz as a test frequency which works well for the last optional part. (Check voltage and frequency with the oscilloscope). The ac base voltage of Q_1 is V_{in}; it is shown as 10 mV$_{pp}$ because of the input voltage divider. Measure the ac signal voltage at the amplifier's output ($V_{out(Q2)}$) and record this value on the last line of Table 6-11. Then use V_{in} and V_{out} to find the measured overall gain, A_v'. Record the measured overall gain in the first line of Table 6-11.

Table 6-10

AC Parameter	Computed Value
$r_{e\,(Q1)}'$	
$r_{e\,(Q2)}'$	
$R_{out(Q1)}$	
$R_{in(Q2)}$	
$A_{v\,(NL)(Q1)}$	
$A_{v(NL)(Q2)}$	

Table 6-11

AC Parameter	Computed Value	Measured Value
A_v'		
$R_{in(Q1)}$		
$R_{out(Q2)}$		
$V_{in(Q1)}$	10 mV	
$V_{out(Q2)}$		

6. The measurement of the total input resistance, $R_{in(tot)}$, is done indirectly by the method discussed in Part 1 using a variable test resistor. The method is repeated here for reference but you need to use a larger test resistor because of the higher input resistance of this amplifier. The test circuit is shown in Figure 6-6. The output signal (V_{out}) is set by V_s to a convenient level with the amplifier operating normally (no clipping or distortion). The output is observed (with an oscilloscope) and the amplitude noted. The resistance of R_{test} is increased until V_{out} drops to one-half the value prior to inserting R_{test}. With this condition, $V_{in} = V_{test}$ and $R_{in(tot)}$

59

must be equal to R_{test}. R_{test} can then be removed and measured with an ohmmeter. The total input resistance, $R_{in(tot)}$, is the same as the ac input resistance to Q_1 ($R_{in(Q1)}$). Using this method, measure $R_{in(Q1)}$ and record the result in Table 6-11.

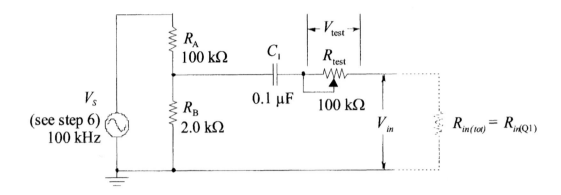

Figure 6-6 Measurement of $R_{in(tot)}$.

7. Measure the output resistance of the amplifier. The computed output resistance is the same as R_{C2}, the load resistor of Q_2. You can measure the output resistance of any amplifier by measuring the loading effect caused by adding a load resistor. It is not necessary that the load resistor be equal to the output resistance. Consider the model of an amplifier shown in Figure 6-5. Assume that you want to indirectly measure R_{out2}. Think of the amplifier as a Thevenin source with the Thevenin resistance represented by the output resistance. You can find this resistance by measuring the unloaded output voltage and the new output voltage when a known load resistor is placed on the output. Use this idea to develop the equation for the output resistance of the amplifier. Then make the measurements and solve for the output resistance.

Optional AGC Circuit:
8. Add the Automatic Gain Control (AGC) shown in Figure 6-7. The AGC circuit is in the box and consists of transistor Q_3, diode D_1, capacitors C_2 and C_6 and resistors R_5 and R_6. C_2, which served as a bypass capacitor in Figure 6-4, has been moved and is now a coupling capacitor for the AGC circuit. This AGC will limit the gain moderately because control is applied to only one stage, whereas in many applications, the gain control voltage is applied to several stages. Test the amplifier with the input signals listed in Table 6-12. Measure and record the output for each input level. Use the ratio of the signals to calculate the gain and record it in the last column of Table 6-12. The input signal is computed from the function generator setting and assumed to be 2% of the function generator value due to the input voltage divider. Graph the overall gain versus the input signal amplitude with AGC on Plot 6-1. Label your graph.

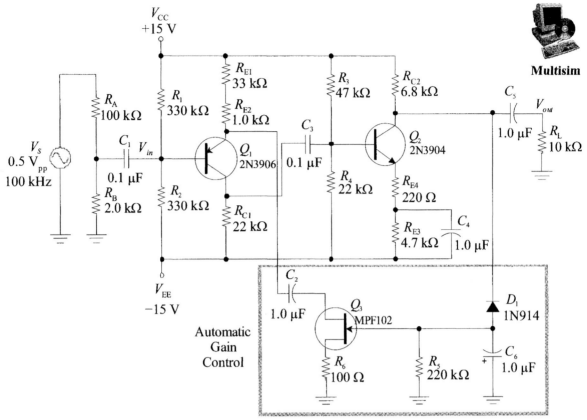

Figure 6-7

Table 6-12

Generator Setting	V_{in}	V_{out}	A_v
0.5 V	10 mV		
1.0 V	20 mV		
2.0 V	40 mV		
4.0 V	80 mV		
6.0 V	120 mV		
8.0 V	160 mV		
10.0 V	200 mV		
20.0 V	400 mV		

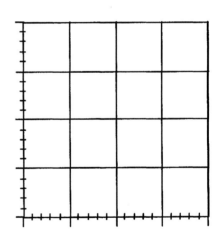

Plot 6-1

Conclusion: Part 3

Questions: Part 3

1. Explain why it is necessary to measure the input resistance indirectly.

2. When you calculated the ac parameters, you were instructed to consider the capacitors as *shorts*. Under what conditions is this assumption *not* warranted?

3. For the circuit in the experiment, the output resistance of both transistors was considered to be the individual collector resistor. Explain.

4. Assume you wanted to use base bias for Q_2 as shown in the circuit of Figure 6-4.
    (a) Explain the changes you would make to the amplifier to accomplish this.

    (b) What disadvantage would result from this change?

5. What is the phase between the input and output signal? Explain your answer.

Multisim Simulation

Multisim

Multisim files for the lab manual are on the website www.prenhall.com/floyd. Open the Multisim file Experiment_06 Multistage-nf. Set up the scope and measure the gain. Compare it with your experimental result.

Notice that the FET AGC control shown in Figure 6-7 is ready to be connected. The MPF102 is not available in Multisim, so a similar transistor, the 2N5485, is used as a replacement. Disconnect C_2 and reverse the orientation; then hook up the AGC circuit as shown in Figure 6-7. Measure the gain, then double the input voltage to 90 mV$_{pp}$ and observe what happens. When you change the input, you will need to let Multisim settle to the final value – it will take a minute or so. Try other values of input voltage and observe the effect of the AGC.

Experiment 7 Power Amplifiers

Power amplifiers are amplifiers that must be able to provide power to drive a load – anywhere from a few watts to hundreds of watts. Heat dissipation and efficiency are always considerations with power amplifiers.

Except in low-power applications, the class-A amplifier is not widely used because it is not particularly efficient. In Part 1 of this experiment, you will measure the power gain and efficiency of a simple class-A power amplifier. The amplifier you will test uses a Darlington arrangement of transistors to improve the power gain. The total power is low to avoid heat, which is associated with power amplifiers.

In Part 2, you will construct a class-B (push-pull) amplifier. The class-B amplifier is much more efficient than the class A amplifier in part 1. Crossover distortion is observed, and then it is eliminated with a circuit called a *diode current-mirror* to bias the transistors into slight conduction. The forward-biased diodes will each have approximately the same 0.7 V drop as the base-emitter junction. The circuit is modified to one that is similar to the Application Activity in the text but uses lower power to avoid heating problems and is simplified to make your lab time efficient.

Reading

Floyd, *Electronic Devices*, Eighth Edition, Chapter 7

Key Objectives

Part 1: Calculate and measure the dc and ac characteristics for a class-A amplifier.
Part 2: Calculate and measure the dc and ac characteristics for a class-B amplifier.

Components Needed

Part 1: The Class-A Power Amplifier

Resistors: one 22 Ω, 2 W, one 10 kΩ, one 22 kΩ
Transistors: one 2N3904, one 2N3053 with heat sink
Capacitors: one 0.22 μF, one 100 μF
One small 8 or 16 Ω speaker

Part 2: The Class-B Power Amplifier

Resistors: one 330 Ω, one 2.7 kΩ, two 10 kΩ, one 68 kΩ
One 1.0 μF capacitor
Transistors: one 2N3906 *pnp*, two 2N3904 *npn* (or equivalent)
Two 1N914 diodes (or equivalent)
One 5 kΩ potentiometer

Part 1: The Class-A Power Amplifier

Caution: **The power transistor can become hot in this experiment.**

1. Measure and record the values of the resistors listed in Table 7-1. R_E is a 2 W resistor. You should not attempt to force its leads into a standard protoboard.

Table 7-1

Resistor	Listed Value	Measured Value
R_1	10 kΩ	
R_2	22 kΩ	
R_E	22 Ω, 2 W	

2. Figure 7-1 shows CC power amplifier using a Darlington arrangement (discussed in Chapter 6 of the text). The input is a 2.5 V_{pp} signal from a generator.

   Compute the parameters listed in Table 7-2. You can treat the transistors as if they were one "super β" transistor but with a 1.4 V base-emitter drop. The unloaded gain ($A_{v(NL)}$) is found by assuming the output coupling capacitor, C_2, is open. The full load gain ($A_{v(FL)}$) can be estimated by assuming the speaker is a resistive load. The r_e' for Q_1 and Q_2 both affect the gain; to simplify the calculation you can estimate the combined effects of r_e' as about 1 Ω in the emitter circuit (shown in the table). Enter your computed values in Table 7-2.

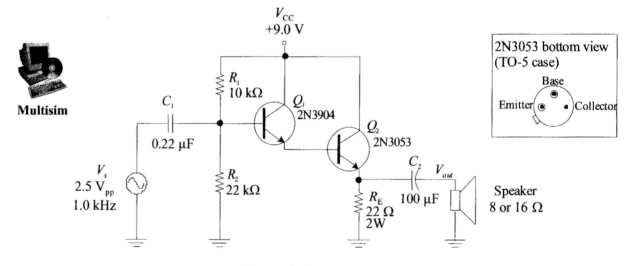

Figure 7-1

Table 7-2

CC Amp ($Q_{1,2}$)	Computed Value	Measured Value
V_B		
V_E		
I_E		
V_{CE}		
r_e'	1 Ω (est.)	
$A_{v(NL)}$		
$A_{v(FL)}$		

3. Construct the circuit shown in Figure 7-1. Measure and record the parameters that you computed in step 2 in Table 7-2. When measuring voltage gain, it is important that no distortion can be observed on the output signal. For the unloaded gain, open C_2 and compare the input from the generator and the voltage across the emitter. The measured and computed values should agree within 10%.

4. In this step, you will determine the power input, load power, and power gain of the amplifier. Enter all values in Table 7-3. The steps are:
    (a) Enter the speaker's resistance, R_L, in Table 7-3. The value should be marked on the speaker.
    (b) Compute the input resistance, R_{in}. The input resistance to the base of Q_2 is much larger than the bias resistors due to the Darlington arrangement. To simplify the calculation, use only the bias resistors in parallel for R_{in}.
    (c) While observing the peak-to-peak output voltage on the speaker, increase the input signal to the onset of clipping. Then measure the maximum undistorted output voltage, V_{out}. Convert the reading to rms voltage.
    (d) Without changing the generator, measure the rms input voltage, V_{in}.
    (e) Compute the power to the load, P_L, from the equation, $P_L = V_{out}^2/R_L$.
    (f) Compute the input power, P_{in}, from the equation, $P_{in} = V_{in}^2/R_{in}$.
    (g) Compute the power gain, A_p, for the circuit from the ratio of the load power to the input power.

Table 7-3

Quantity	Value
Load resistance, R_L	
Input resistance, R_{in}	
Output rms voltage, V_{out}	
Input rms voltage, V_{in}	
Load power, P_L	
Input power, P_{in}	
Power gain, A_p	

Table 7-4

Quantity	Value
Quiescent power, P_Q	
Load power, P_L	
Efficiency (percentage)	

5. The approximate efficiency of the amplifier is the ratio of the power delivered to the load divided by the dc power delivered by the power supply. The power from the supply is essentially the same as the quiescent power, P_Q, required to maintain bias and set up the operating conditions. Calculate P_Q as the product of I_E and V_{CC}. (Ignore the very small current in the bias resistors and the contribution to load power with the signal). The load power is the same as entered in Table 7-3. Calculate the ratio of load power to quiescent power as the approximate efficiency (converted to a percentage). Enter results from this step in Table 7-4.

Conclusion: Part 1

Questions: Part 1

1.   (a) Would the efficiency of the amplifier be higher if the emitter resistor were replaced with the load resistor (speaker)? What is the disadvantage?

    (b) Would the efficiency of the amplifier be higher if the input signal is larger? Explain your answer.

2.   What is the advantage of the Darlington arrangement for the CC amplifier in this experiment?

Part 2: The Class-B Power Amplifier

1.   Measure and record the values of the resistors listed in Table 7-5.

2.   Construct the push-pull amplifier shown in Figure 7-2. Set the generator for a 10 V_{pp} sine wave at 1.0 kHz. Be sure that there is no dc offset from the generator. The dual positive and negative power supplies offer the advantage of not requiring coupling capacitors, provided the signal does not have a dc offset.

Table 7-5

Resistor	Listed Value	Measured Value
R_L	330 Ω	
R_1	10 kΩ	
R_2	10 kΩ	
R_3	68 kΩ	
R_4	2.7 kΩ	

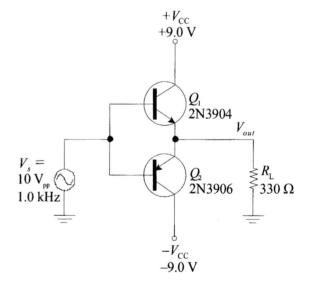

Figure 7-2

66

3. Sketch the input and output waveforms you observe on Plot 7-1, which represents the scope face. Show the amplitude difference between the peak input waveform and the output waveform and note the crossover distortion on your plot. Label your plot for voltage and time and add a title.

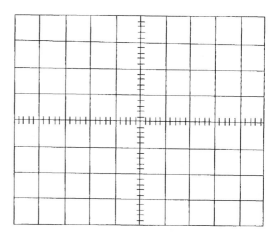

Plot 7-1

4. Add the diode current mirror bias shown in Figure 7-3. Compute and record the dc parameters listed in Table 7-6. The dc emitter voltage will be 0 V if each half of the circuit is identical and there is no dc offset from the generator. The current in R_1 can be found by applying Ohm's law. This current is nearly identical to I_{CQ} because of current-mirror action.

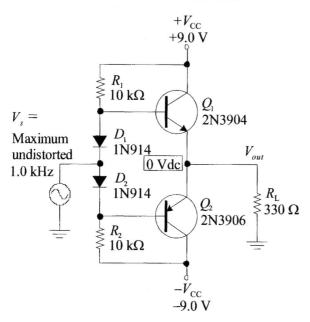

Figure 7-3

Table 7-6

DC Parameter	Computed Value	Measured Value
V_E		
V_{B1}		
V_{B2}		
$I_{R1} = I_{CQ}$		

Table 7-7

AC Parameter	Computed Value	Measured Value
$V_{p(out)}$		
$I_{p(out)}$		
$P_{(out)}$		

5. Compute and record the ac peak parameters listed in Table 7-7. Compute the maximum undistorted output voltage (V_p) and current (I_p) based on the measured load resistance. With dual-power supplies, the output can swing to a maximum positive value that is about 1.5 V less than V_{CC}. Then compute the output power. The power is found by $P_{out} = 0.5I_{p(out)}V_{p(out)}$. By substituting for I_p, the power out can also be expressed as:

$$P_{out} = \frac{V_{p(out)}^2}{2R_L}$$

67

6. With the signal generator off, apply power. Measure and record the dc parameters listed in Table 7-6.

7. Turn on the signal generator and ensure that there is no dc offset. While viewing V_{out}, adjust the generator for the maximum unclipped output voltage. Measure this voltage and enter it as $V_{p(out)}$ in Table 7-7.

Application Activity

8. Another method for driving a push-pull amplifier is shown in Figure 7-4. This is essentially the method used in the Application Activity in the text but with lower power. The signal is first amplified by Q_3, a common-emitter amplifier. The quiescent current in the collector circuit produces the same dc conditions as in the circuit of Figure 7-3. The bias adjust allows the dc output voltage to be set to zero to compensate for tolerance variations in the components. Compute and record the dc parameters listed in Table 7-8. Assume the bias potentiometer is set to 3 kΩ and apply the voltage-divider rule to find $V_{B(Q3)}$. Note that the voltage across the divider string is the difference between $+V_{CC}$ and $-V_{CC}$.

Table 7-8

DC Parameter	Computed Value	Measured Value
$V_{B(Q3)}$		
$V_{E(Q3)}$		
I_{CQ3}		

Table 7-9

AC Parameter	Computed Value	Measured Value
$A_v'(Q3)$		

9. Compute and record in Table 7-9 the voltage gain of Q_3. The gain of Q_3 is the load presented to the collector circuit of Q_3 divided by the resistance of the Q_3 emitter circuit. The voltage gain of the push-pull amplifier is nearly 1.0, so the total voltage gain of the amplifier, A_v', is approximately equal to the gain of Q_3. That is,

$$A_v' \cong A_{v(Q3)} \cong \frac{R_1 || \{\beta_{Q1}(R_L + r'_{e(Q1)})\}}{r'_{e(Q3)} + R_4}$$

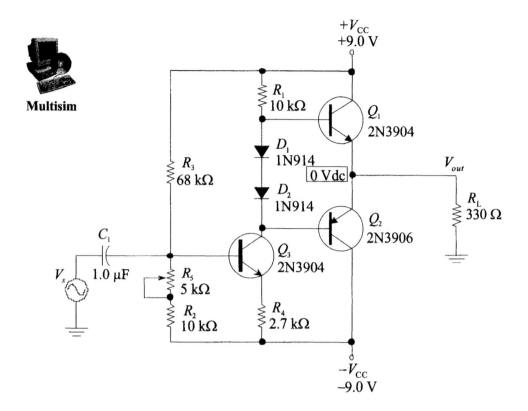

Figure 7-4

10. Connect the circuit shown in Figure 7-4. Measure the dc voltage across the load resistor and adjust the bias potentiometer (R_5) for 0 V at the output. Measure and record the voltages listed in Table 7-8.

11. Set the signal generator for the maximum unclipped voltage across the load resistor and measure the total voltage gain, A_v'. Enter the measured gain in Table 7-9.

Conclusion: Part 2

Questions: Part 2
1. Assume the circuit in Figure 7-3 has a positive half-wave rectified output. What are three failure(s) that could account for this?

69

2.   If one of the diodes in Figure 7-3 opens, what symptoms will it produce?

3.   The total voltage gain was fairly low for the circuit of Figure 7-4. What change to the circuit would you suggest to increase the voltage gain?

Multisim Simulation

Multisim

Multisim files for the lab manual are on the website www.prenhall.com/floyd. Open the Multisim file Experiment_07_Class_A-nf. This is the circuit in Figure 7-1, but the speaker has been replaced in the simulation with an 8 Ω load resistor. Connect the scope and measure the rms voltage to the load. Compare the power delivered to the load to the power supplied from the power supply and calculate the efficiency. Compare the simulation results with your results in Part 1. Then open the file Experiment_07_Class_A-fault. Isolate the problem using the dc voltmeter and/or the oscilloscope.

For Part 2, the Class-B amplifier in Figure 7-4 is simulated in Multisim. Open the file Experiment_07_Push-Pull-nf. Connect a dc voltmeter to the output and adjust R_5 for minimum dc output. Observe the output using the oscilloscope. Determine the overall gain and compare it to your experimental result. Then open the file Experiment_07_Push-Pull-fault. Isolate the problem using the dc voltmeter and/or the oscilloscope.

Experiment 8 Field-Effect Transistors

Field-effect transistors (FETs) work on an entirely different principle than
BJTs. They are voltage-controlled devices rather than current-controlled
devices as in the case of BJTs. The first part of this experiment investigates
the characteristic curve for a FET, which is a plot of drain source voltage as
a function of drain current for a given gate source voltage.

Part 2 introduces a circuit that uses the ohmic region of the JFET as a
variable resistor to control the gain of a basic BJT amplifier. The ohmic
region is near $V_{DS} = 0$, and in fact for the circuit you will test, the operating
point is set at $V_{DS} = 0$ by connecting a capacitor in series with the drain.
After testing the basic idea, the circuit is modified to illustrate automatic
gain control (AGC).

In Part 3, you will test a JFET circuit that is set up to amplify small
voltages from a high impedance transducer such as the one discussed in the
Application Activity of the text. This particular amplifier will work with
small positive or negative voltages but the FET remains biased with negative
gate to source voltage for these inputs. The results are similar to the dual
gate D-MOSFET in the text as you will see.

Reading
Floyd, *Electronic Devices*, Eighth Edition, Chapter 8

Key Objectives
Part 1: Measure and plot the characteristic curves for an *n*-channel JFET.
Part 2: Construct and test a circuit that uses a JFET as a voltage-controlled
resistor. Then modify the circuit for automatic gain control.
Part 3: Test a circuit in which a dc voltage from a simulated transducer is
amplified.

Components Needed
Part 1: JFET Characteristic Curve
Resistors: one 100 Ω, one 10 kΩ
One 2N5458 *n*-channel JFET (or equivalent)
One LED
One milliammeter $I_{FS} = 10$ mA. (I_{FS} is the full-scale reading.)

Part 2: JFET as a Voltage-Controlled Resistor
Resistors: one of each: 3.9 kΩ, 6.2 kΩ, 39 kΩ, 56 kΩ, 100 kΩ, 5.1 MΩ
One 2N3904 *npn* transistor
One 2N5458 *n*-channel JFET
One signal diode, 1N914
Capacitors: two 1.0 μF, two 10 μF

Part 3: JFET as a DC Amplifier
Resistors: one of each: 180 Ω, 620 Ω, 3.9 kΩ, 5.6 kΩ, 1.0 MΩ
One 2N5458 *n*-channel JFET
One 10 kΩ potentiometer

Part 1: JFET Characteristic Curve
1. Measure and record the values of the resistors listed in Table 8-1. R_1 is used for protection in case the JFET is forward-biased accidentally. R_2 serves as a current-sensing resistor.

2. Construct the circuit shown in Figure 8-1. Start with V_{GG} and V_{DD} at 0 V. Connect a voltmeter between the drain and source. Keep V_{GG} at 0 V and slowly increase V_{DD} until V_{GS} is 1.0 V. (V_{DS} is the voltage between the transistor's drain and source.)

Table 8-1

Resistor	Listed Value	Measured Value
R_1	10 kΩ	
R_2	100 Ω	

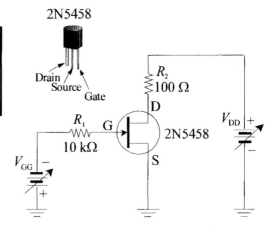

Figure 8-1

3. With V_{DS} at 1.0 V, measure the voltage across R_2 (V_{R2}). Compute the drain current, I_D, by applying Ohm's law to R_2. Note that the current in R_2 is the same as I_D for the transistor. Use the measured voltage, V_{R2}, and the measured resistance, R_2, to determine I_D. Enter the measured value of V_{R2} and the computed I_D in Table 8-2 under the columns labeled $V_G = 0$ V.

Table 8-2

V_{DS}	$V_G = 0$ V		$V_G = -0.5$ V		$V_G = -1.0$ V		$V_G = -1.5$ V	
	V_{R2}	I_D	V_{R2}	I_D	V_{R2}	I_D	V_{R2}	I_D
1.0 V								
2.0 V								
3.0 V								
4.0 V								
6.0 V								
8.0 V								

4. Without disturbing the setting of V_{GG}, slowly increase V_{DD} until V_{DS} is 2.0 V. Then measure and record V_{R2} for this setting. Compute I_D as before and enter the measured voltage and computed current in Table 8-2 under the columns labeled $V_G = 0$ V.

5. Repeat step 4 for each value of V_{DS} listed in Table 8-2.

6. Adjust V_{GG} for -0.5 V. This applies -0.5 V between the gate and source because there is almost no gate current into the JFET and almost no voltage drop across R_1. Reset V_{DD} until $V_{DS} = 1.0$ V. Measure V_{R2} and compute I_D as before. Enter the values in Table 8-2 under the columns labeled $V_G = -0.5$ V.

7. Without changing the setting of V_{GG}, adjust V_{DD} for each value of V_{DS} listed in Table 8-2 as before. Compute the drain current at each setting, and enter the voltage and current values in Table 8-2 under the columns labeled $V_G = -0.5$ V.

8. Adjust V_{GG} for -1.0 V. Reset V_{DD} until $V_{DS} = 1.0$ V. Measure V_{R2} and calculate I_D. Repeat step 7, entering the data in the columns labeled $V_G = -1.0$ V.

9. Adjust V_{GG} for -1.5 V. Reset V_{DD} until $V_{DS} = 1.0$ V. Measure V_{R2} and calculate I_D. Repeat step 7, entering the data in the columns labeled $V_G = -1.5$ V.

10. The data in Table 8-2 represent four drain characteristic curves for your JFET. The drain characteristic curve is a graph of V_{DS} versus I_D for a constant gate voltage. Plot the four drain characteristic curves on Plot 8-1. Choose a scale for I_D that allows the largest current observed to fit on the graph. Label each curve with the gate voltage it represents and add an appropriate title to the graph.

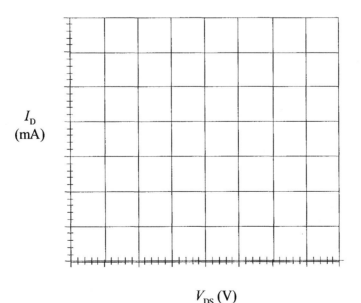

I_D
(mA)

V_{DS} (V)

Plot 8-1

11. Set V_{DD} for +12 V and V_{GG} for 0 V. Monitor the voltage across R_2 and slowly increase the negative gate voltage. When the voltage across R_2 reaches zero, note the gate voltage. Record this value in Table 8-3 as $V_{GS(off)}$. Record I_{DSS} by reading Plot 8-1. These are the parameters for your JFET that will be used in Part 2.

Table 8-3

Measured JFET Parameters	
$V_{GS(off)} =$	
$I_{DSS} =$	

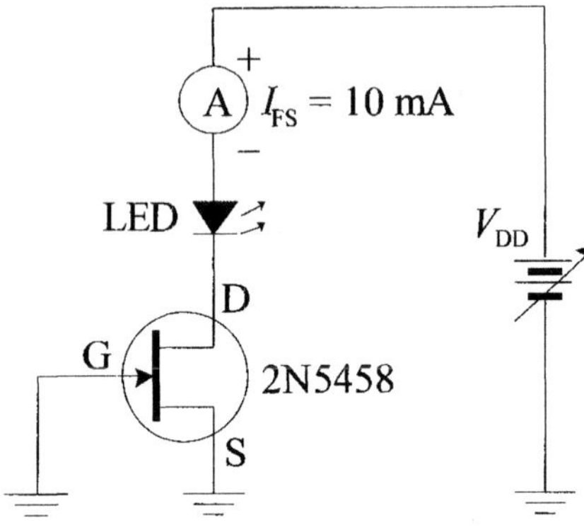

Figure 8-2

12. Construct the circuit shown in Figure 8-2. This circuit is a JFET connected as a two-terminal constant-current source. Monitor the drain voltage while you increase the drain power supply from 0 V to +15 V. Notice the drain voltage where constant-current begins. This point is imprecise but about the same as the absolute value of $V_{GS(off)}$. Compare the ammeter reading with the maximum current found when you tested the FET in steps 3, 4 and 5. Observations:

_____

Conclusion: Part 1

Questions: Part 1

1. Explain how you could develop the transconductance curve from the data taken in this experiment.

2. (a) Does the experimental data indicate that the transconductance is constant at all points?

    (b) From your experimental data, what evidence indicates that a JFET is a nonlinear device?

Part 2: JFET as a Voltage-Controlled Resistor
Voltage Control of Gain
1. Measure and record the values of the resistors listed in Table 8-4.

Table 8-4

Resistor	Listed Value	Measured Value
R_1	56 kΩ	
R_2	39 kΩ	
R_E	6.2 kΩ	
R_C	3.9 kΩ	
R_3	100 kΩ	

Table 8-5

DC Quantity	Computed Value	Measured Value
V_B		
V_E		
I_E		
V_C		
V_{CE}		

2. As discussed in the text in Section 8-4, a JFET can function as a voltage-controlled resistor, which is useful in a variety of circuits. Figure 8-3 shows a BJT amplifier with a JFET used for gain control. The JFET is operating in the ohmic region and provides an ac resistance path to ground in the emitter circuit. To analyze the circuit, start with the dc values. Calculate and record the dc quantities listed in Table 8-5. The simplified model is valid for calculating the bias voltage ($\beta R_E \geq 10R_2$).

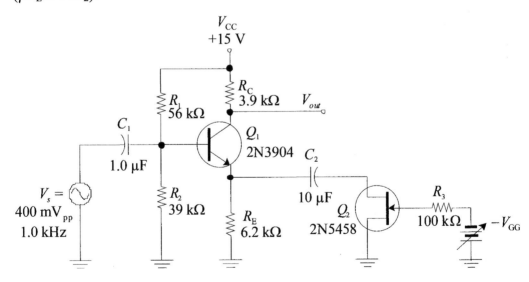

Figure 8-3

3. Construct the amplifier shown in Figure 8-3. The signal generator should be turned off and V_{GG} is set to a minimum (0 V). Measure and record the dc voltages listed in Table 8-5.

4. The ac emitter resistance is r_d in parallel with R_E. At $V_{GS} = 0$, $r_d = r_0$. You can estimate the maximum gain if you know the channel resistance at $V_{GS} = 0$. At $V_{GS} = 0$, the channel resistance is approximately

$$r_0 \cong -\frac{V_{GS(off)}}{2I_{DSS}}$$

If you are using the same transistor as in Part 1, you can estimate the channel resistance at $V_{GS} = 0$ if you use the values of $V_{GS(off)}$ and I_{DSS} from Table 8-3. Use the JFET channel resistance, r_0, to estimate the maximum gain of the BJT amplifier. Enter your prediction in Table 8-6.

If you do not have the data for the FET used in this part, an estimate of the channel resistance for a typical 2N5458 transistor can be made by using the manufacturer's typical values. Unfortunately, JFETs have a very wide range of specifications even among identical types, so this result must be considered approximate. From the manufacturer's specified values,

$$r_0 = -\frac{-3.5\ \text{V}}{2(6\ \text{mA})} = 292\ \Omega$$

Table 8-6

	Predicted Gain	Measured V_{out}	Measured Gain
Maximum Gain $V_{GG} = 0$ V			
Gain with $V_{GG} = -0.5$ V			
Gain with $V_{GG} = -1.0$ V			
Gain with $V_{GG} = -1.5$ V			
Gain with $V_{GG} = -2.0$ V			
Gain with $V_{GG} = -2.5$ V			
Gain with $V_{GG} = -3.0$ V			
Gain with $V_{GG} = -5.0$ V			

5. Set your function generator for a signal level of 400 mV$_{pp}$ at 1.0 kHz. Adjust V_{GG} to each voltage in Table 8-6 and measure V_{out}. Use the measured output voltage to determine the measured gain. Enter your results in Table 8-6.
    Observations:

_____

Automatic Gain Control (AGC)
6. The amplifier you constructed illustrates how voltage control can change the gain. In a practical application, the gain is controlled by the input signal. When the signal is large, the JFET will tend to be biased off, which increases the effective emitter resistance and reduces the gain. The idea is shown in Figure 8-4. The negative excursions of the signal are used to determine the amount of gate voltage, and hence the gain.

Change the circuit to the one shown in Figure 8-4. Notice the polarity of C_4. Adjust the input signal to 100 mV$_{pp}$ and measure the output voltage. Use the output voltage to calculate the gain for each of the input signals listed in Table 8-7. Notice that the last two values would significantly overdrive the amplifier if it were not for the gain control.

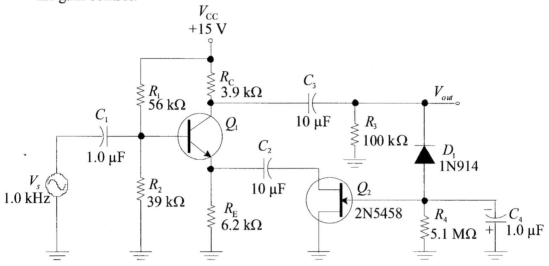

Figure 8-4

Table 8-7

	Measured V_{out}	Measured Gain
$V_{in} = 100$ mV$_{pp}$		
$V_{in} = 400$ mV$_{pp}$		
$V_{in} = 800$ mV$_{pp}$		
$V_{in} = 1.2$ V$_{pp}$		

Conclusion: Part 2

Questions: Part 2

1. How does the automatic gain control for the circuit in Figure 8-4 generate a negative voltage to the gate of the FET?

2. What is the purpose of R_4 and C_4 for the circuit in Figure 8-4?

Part 3: JFET as a DC Amplifier

1. Measure and record the values of the resistors listed in Table 8-8.

Table 8-8

Resistor	Listed Value	Measured Value
R_1	5.6 kΩ	
R_2	620 Ω	
R_3	1.0 MΩ	
R_S	180 Ω	
R_D	3.9 kΩ	

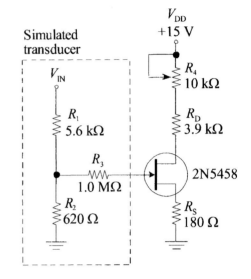

Figure 8-5

2. Construct the circuit shown in Figure 8-5. R_1 and R_2 serve as an input voltage-divider to make it easier to set the small voltages from the simulated transducer. R_3 represents the high source resistance, typical of many transducers. V_{IN} is a separate DC source that can be adjusted for positive or negative voltages. (*Note:* For negative voltages, if you have an isolated supply, you can use the positive output as common and use the common as V_{IN}. Check with your instructor if you are not sure.)

3. Set up V_{IN} as a positive supply voltage and adjust it for +300 mV at the gate. Then adjust R_4 for 3.0 V at the drain. The source voltage should be greater than +300 mV at this setting to assure that the FET is reverse-biased. Record the drain voltage in the first row of Table 8-9.

4. Without adjusting R_4, reduce V_{IN} to +200 mV as measured at the gate. Measure the drain voltage, and record it in the next line of Table 8-9.

5. Repeat step 4 for the other input values listed on the table. The power supply leads will need to be reversed for the negative input values.

Table 8-9

V_{IN}	V_D (measured)
300 mV	
200 mV	
100 mV	
0 V	
−100 mV	
−200 mV	
−300 mV	
−400 mV	

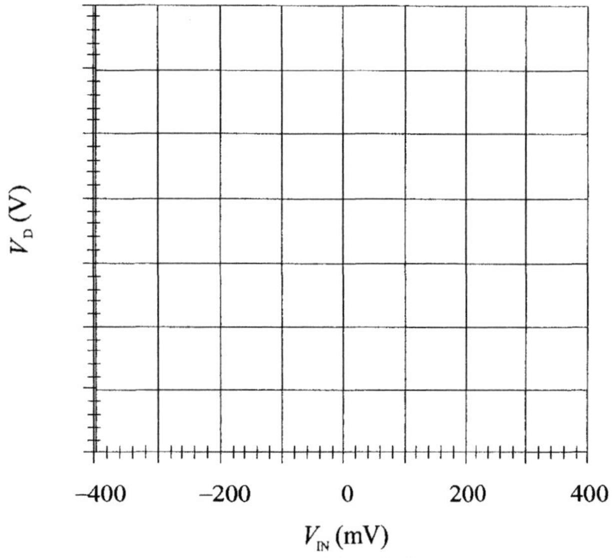

Plot 8-2

6. Plot the data from Table 8-9 in Plot 8-2. You will need to add labels to the *y*-axis.

Conclusion: Part 3

Questions: Part 3

1. For the limited range of the input voltage, is the output linear? Explain your answer.

2. Explain how the FET is able to have a negative-biased gate even with positive input voltages.

3. How would you expect the data to be affected if the power supply voltage is raised?

Multisim Simulation

Multisim

Multisim files for the lab manual are on the website www.prenhall.com/floyd. Open the Multisim file Experiment_08_FET_AGC-nf. This circuit is similar to the AGC circuit in Figure 8-4. The FET is used as a voltage controlled resistor which in turn, controls the gain. This circuit has a longer time constant than the AGC circuit given earlier in Experiment 6 (see Figure 6-7) because of the lower frequency. The input signal is shown as 35.4 mV, which is equivalent to 100 mV_{pp} as a starting point. The computer simulation requires a number of iterations to settle. You can observe the automatic gain control as the input signal amplitude is changed. (Stop the simulation; change the amplitude and restart it). Notice that a BF545A transistor is used in the simulation because Multisim does not have a 2N5458 transistor in its library.

Experiment 9 FET Amplifiers and Switching Circuits

> **F**ield-effect transistors give designers important advantages for linear
> amplifiers. This experiment focuses on two important types – the common
> source amplifier and the common-drain amplifier.
>
> In Part 1, you will investigate a common-source amplifier using self-
> bias and consider several troubleshooting problems.
>
> In Part 2, a common-drain amplifier is constructed and tested. After
> testing a self-biased amplifier, the circuit is changed to a current-source
> biased amplifier using another JFET for the current source.
>
> In Part 3, a cascode amplifier is investigated that is similar to the
> Application Activity. High gain radio-frequency amplifiers are not suited to
> laboratory protoboards, so the circuit has been changed to use general
> purpose JFETs in a modified amplifier operating at 500 kHz. Even so, you
> will need to keep leads short and pay attention to grounds to avoid noise
> problems.

Reading

Floyd, *Electronic Devices*, Eighth Edition, Chapter 9

Key Objectives

Part 1: Calculate and measure dc and ac parameters for a common-source
amplifier.

Part 2: Calculate and measure dc and ac parameters for two common-drain
amplifiers.

Part 3: Measure parameters for a cascode amplifier constructed from two JFETs.

Components Needed

Part 1: The Common-Source JFET Amplifier

Resistors (one of each): 620 Ω, 1.0 kΩ, 3.3 kΩ, 10 kΩ, 100 kΩ, 1.0 MΩ
One 2N5458 *n*-channel JFET
Capacitors (one if each): 0.1 µF, 1.0 µF, 10 µF

Part 2: The Common-Drain JFET Amplifier

Resistors (one of each): 470 Ω, 1.0 kΩ, 10 kΩ, 100 kΩ, 1.0 MΩ
Two 2N5458 *n*-channel JFET
One 1 kΩ potentiometer
Capacitors: one 0.1 µF, one 1.0 µF, one 10 µF

Part 3: A Cascode Amplifier

Resistors: one 100 Ω, one 620 Ω, one 5.6 kΩ, two 1.0 MΩ, one 1.5 MΩ
Capacitors: four 0.1 µF
Transistors: two 2N5458 *n*-channel JFETs

Part 1: The Common-Source JFET Amplifier

1. Measure and record the values of the resistors listed in Table 9-1.

2. Construct the common-source (CS) amplifier shown in Figure 9-1. Set the signal generator for a 500 mV$_{pp}$ sine wave at 1.0 kHz. Check the amplitude and frequency with your oscilloscope.

Table 9-1

Resistor	Listed Value	Measured Value
R_S	1.0 kΩ	
R_D	3.3 kΩ	
R_G	1.0 MΩ	
R_L	10 kΩ	

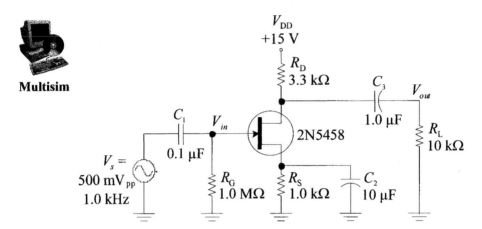

Figure 9-1

3. Measure the dc voltages listed in Table 9-2 and compute I_D. Set the function generator for a 500 mV$_{pp}$ sine wave and measure the ac quantities listed. Compare the input and output ac voltage by viewing V_{in} and V_{out} simultaneously. Measure the voltage gain and note the phase difference (0° or 180°) between the input and output signal. Enter all data from this step in Table 9-2.

Table 9-2 Parameters for CS Amplifier

Quantity	DC values	AC values
Gate voltage, V_G		
Source voltage, V_S		
Drain voltage, V_D		
Drain current, I_D		
Input voltage, V_{in}		
Output voltage, V_{out}		
Voltage gain, A_v		
Phase difference		

4. Change the source resistor from 1.0 kΩ to a smaller value. If you are using the same JFET as in Experiment 8, you can calculate the best self-bias resistor using the method described in Example 8-9 of the text. Otherwise, use a 620 Ω resistor for R_S. You should observe a slight increase in gain with the smaller resistor, despite the fact that it is bypassed. Can you explain this gain increase? (*Hint:* consider g_m.)
   Observation:

   _____

5. Now change the load resistor from 10 kΩ to 100 kΩ. Does the gain change? Explain your observation:

   _____

Troubleshooting
6. Assume each of the faults listed in Table 9-3 is in the common-source amplifier circuit of Figure 9-1. Predict the outcome with the fault in place. Then put the fault in the circuit and test your prediction.

   Table 9-3

Fault	Observation
C_2 is open	
Source and drain reversed	
V_{DD} drops to +12 V	
R_G open	

Conclusion: Part 1

Questions: Part 1
1. If the operating point is changed in the circuit of Figure 9-1 because of a different source resistor, is there any effect on the input or output impedance? Explain.

2. Compare the amplifier in Figure 9-1 with the CE amplifier in Experiment 6 (Figure 6-1). Which is better suited to amplify a signal from a source that has a 100 kΩ Thevenin resistance? Explain your answer.

Part 2: The Common-Drain JFET Amplifier

1. A self-biased common-drain (CD) circuit is shown in Figure 9-2. Connect the circuit and measure the dc voltage at the drain, source, and gate and compute I_D. Observe the input and output ac voltage with the oscilloscope. Measure the voltage gain and note the phase. Enter the data in Table 9-4.

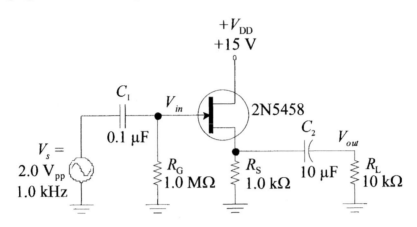

Figure 9-2

Table 9-4 Parameters for CD Self-Biased Amplifier

Quantity	DC values	AC values
Gate voltage, V_G		
Source voltage, V_S		
Drain voltage, V_D		
Drain current, I_D		
Input voltage, V_{in}		
Output voltage, V_{out}		
Voltage gain, A_v		
Phase difference		

Adding Current-Source Bias

2. The voltage gain is much less than 1.0 due to the transconductance, g_m. This can be visualized as an internal resistance equal to the reciprocal of g_m. This internal resistance ($1/g_m$) is analogous to r_e' of a bipolar transistor, but is larger for a given current. It forms a voltage divider with R_S as shown in Figure 9-3. To improve the gain, the JFET current source (with its high internal resistance) can be used.

    Change the previous self-biased circuit to include the current-source biasing shown in Figure 9-4. This circuit does not have coupling capacitors (an advantage for the low-frequency response) but the source resistor now includes a variable resistor to adjust the dc offset. Notice that the output is taken at the drain of Q_2, which is the current source. (The amplifier is still referred to as a *common-drain* amplifier because of Q_1.) Start with R_{S2} set near the center of its resistance.

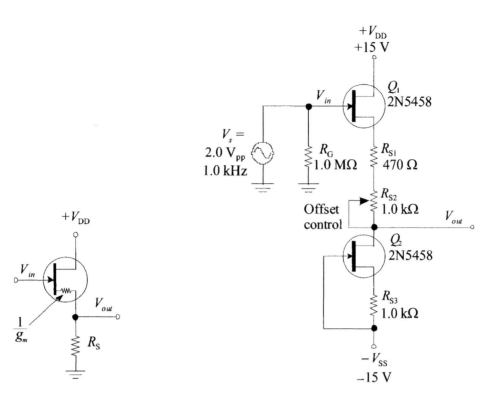

Figure 9-3 $1/g_m$ and R_S form a voltage divider.

Figure 9-4 Q_2 is a current source that has high drain-source resistance.

3. DC couple your oscilloscope and view the output. Adjust R_{S2} for no dc offset in the output. Measure and record the dc and ac quantities listed in Table 9-5 for the common-drain, current-source amplifier. Notice the improvement in the gain over the self-biased common-drain circuit you constructed in step 1.

Table 9-5 Parameters for CD Current-Source Biased Amplifier

Quantity	DC values	AC values
Q_1 gate voltage, V_G		
Q_1 source voltage, V_S		
Q_1 drain voltage, V_D		
Q_2 gate voltage, V_G		
Q_2 source voltage, V_S		
Q_2 drain voltage, V_D		
Drain current, I_D		
Input voltage, V_{in}		
Output voltage, V_{out}		
Voltage gain, A_v		
Phase difference		

4. With no dc offset, the load resistor can be directly connected to the output. Test the effect of a 10 kΩ load on the amplifier. Then test the signal clipping point by increasing the input signal from the generator.
    Observations:

    _____

5. Try switching the two FETs. Does the DC offset need to be readjusted?
    Observations:

    _____

Conclusion: Part 2

Questions: Part 2
1. Why was the gain much better with current-source biasing than self-bias?

2. Estimate the input and output resistance of the amplifier in Figure 9-4 based on the observations you made in step 4.

Application Activity

Part 3: A Cascode Amplifier
1. Measure and record the values of the resistors listed in Table 9-6.

2. Construct the cascode amplifier shown in Figure 9-5. The circuit is similar to the active antenna circuit given in the Application Activity but the output is taken from the source of Q_1 instead of the drain. In this circuit, Q_1 serves as a current load for Q_2. The purpose of R_1 and R_2 is to attenuate the signal from the function generator by a factor of 10, to make it easier to view the small signal. Notice that C_3 is not connected to ground.

Table 9-6

Resistor	Listed Value	Measured Value
R_1	5.6 kΩ	
R_2	620 Ω	
R_3	1.0 MΩ	
R_4	1.5 MΩ	
R_G	1.0 MΩ	
R_S	100 Ω	

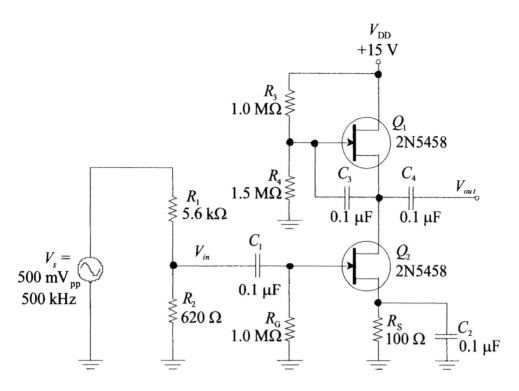

Figure 9-5

3. Turn on the power supply. With the function generator off, measure the dc voltages shown in Table 9-7 and record their values.

4. Set the function generator to a 500 kHz sine wave at 500 mV$_{pp}$ (this is shown in Table 9-8 as 50 mV$_{pp}$ for V_{in} because of the input voltage divider.) You can probe the input at the function generator, making it easier to view the small signal. If you notice noise on the signal, make sure grounds are connected to a single point; you can add a large capacitor across the input power connections if necessary. Measure V_{out} and use it to calculate the gain. Record the data in Table 9-8 for the amplifier.

Table 9-7

Parameter	Measured Value
$V_{G(Q1)}$	
$V_{S(Q1)}$	
$V_{G(Q2)}$	
$V_{S(Q2)}$	

Table 9-8 (Data for 500 kHz)

Parameter	Measured Value
V_{in}	50 mV$_{pp}$
V_{out}	
$A_{v(Q1/Q2)}$	

5. Change the function generator to a frequency of 50 kHz. Measure V_{out} and calculate the gain. Record the data in Table 9-9.

Table 9-8 (Data for 50 kHz)

Parameter	Measured Value
V_{in}	50 mV$_{pp}$
V_{out}	
$A_{v(Q1/Q2)}$	

6. Reduce the power supply voltage while observing the output. What if any effect does the supply voltage have on the output?

Conclusion: Part 3

Questions: Part 3

1. Are both FETs operating with negative gate-to-source voltage? Explain your answer.

2. Why do you think the gain was different in step 6 than in step 5?

Multisim Simulation

Multisim

Multisim files for the lab manual are on the website www.prenhall.com/floyd. Open the Multisim file Experiment_09_CS_amp-nf. This circuit is the same as Figure 9-1 but with a BF545A FET. Compare the simulation with the experimental result. Then open the file Experiment_09_CS_amp-fault. Notice that the signal is distorted at the output. Troubleshoot the circuit and find the problem. You can even fix this problem and check your fix.

Experiment 10 Amplifier Frequency Response

In a typical transistor amplifier, the low-frequency response is determined by the coupling and bypass capacitors. The analysis of the low-frequency response is done by considering each capacitor's discharge path and forming an equivalent (and simplified) *RC* circuit.

      In Part 1, you analyze the low-frequency response of a single-stage BJT amplifier. The response due to each capacitor is then measured by isolating capacitors, one at a time, using a method described in the procedure. You will then calculate the value of capacitance that can be used as a bypass capacitor to raise the lower cutoff frequency to 300 Hz, as was done in the Application Activity. (To simplify the experiment, only a one-stage amplifier is tested. The two-stage amplifier that is in the Chapter 10 Application Activity was previously constructed in Experiment 6, Part 3.)

      In Part 2, the upper critical frequency is investigated. The Miller effect capacitance has the greatest impact on the response of inverting amplifiers. Typically, the internal capacitance between electrodes is important only at frequencies well above 1 MHz. To reduce these frequencies so they can be tested with ordinary lab equipment, additional capacitance will be added between the various leads of the transistor. Again, the key to analysis is the simplified equivalent circuit.

Reading

      Floyd, *Electronic Devices*, Eighth Edition, Chapter 10

Key Objectives

      Part 1: Compute and measure the three lower critical frequencies for a CE amplifier and use them to compute the overall lower critical frequency, f_{cl}; then measure f_{cl}.

      Part 2: Compute and measure the three upper critical frequencies for a CE amplifier and use them to compute the overall upper critical frequency, f_{cu}; then measure f_{cu}.

Components Needed

Part 1: Low-Frequency Response

      Resistors: one 10 Ω, one 47 Ω, one 560 Ω, one 1.0 kΩ, one 3.9 kΩ, two 10 kΩ, one 68 kΩ

      One 2N3904 *npn* transistor

      Capacitors: one 0.22 μF, one 1.0 μF, one 100 μF, two 1000 μF, one to be determined by student

Part 2: High-Frequency Response

      Same as Part 1 plus three 100 pF capacitors

Part 1: Low-Frequency Response

1. Measure and record the values of the resistors listed in Table 10-1. You will use the same resistors in Part 2, so it will not be necessary to measure them again.

Table 10-1

Resistor	Listed Value	Measured Value
R_A	1.0 kΩ	
R_B	47 Ω	
R_1	68 kΩ	
R_2	10 kΩ	
R_{E1}	10 Ω	
R_{E2}	560 Ω	
R_C	3.9 kΩ	
R_L	10 kΩ	

Table 10-2

Parameter	Computed Value	Measured Value
V_B		
V_E		
I_E		
V_C		
V_{CE}		
r_e		
A_v		
V_{out}		

2. Compute the ac and dc parameters listed in Table 10-2 for the CE amplifier shown in Figure 10-1. R_A and R_B are not part of the amplifier but are only included as an input attenuator that will provide a 20 mV$_{pp}$ signal to the input (V_{in}). Tabulate your calculations in Table 10-2 (the first five are dc parameters; the last three are ac parameters).

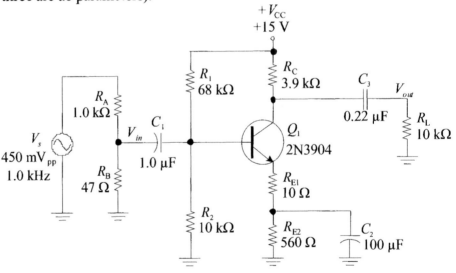

Figure 10-1

3. Construct the amplifier shown in Figure 10-1. Then measure and record the parameters listed in Table 10-2 and confirm your calculations. You can assume the input signal is 20 mV$_{pp}$, which is difficult to measure accurately, if the signal generator is confirmed to be 450 mV$_{pp}$.

4. To compute the low-frequency response, it is necessary to find the equivalent resistance, R_{eq}, that represents the ac charge and discharge path for each capacitor.

Tracing the path for C_1, you see the paths to ground as illustrated in Figure 10-2 with the dotted lines. (Recall that the power supply is an ac ground.) On the right side of C_1 are the bias resistors (R_1 and R_2) and the ac resistance of the emitter circuit consisting of $R_{E1} + r_e$. Together, these resistors are equivalent to a single resistance, R_{in}. On the left side of C_1, are two paths – the series combination of ($R_A + R_{th}$) is in parallel with R_B. Thus, the total equivalent resistance for C_1 is:

$$R_{eq} = R_{in} + (R_A + R_{th}) \parallel R_B$$
$$= \beta_{ac}(R_{E1} + r_e) \parallel R_1 \parallel R_2 + (R_A + R_{th}) \parallel R_B$$

Using this equation, compute the equivalent resistance seen by C_1. It is useful if you know the β_{ac} for your transistor; if you do not know it, you can assume a typical value of 200. Enter the computed value in Table 10-3.

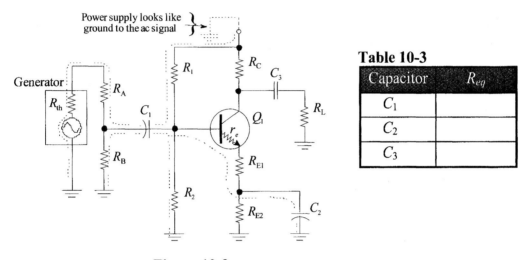

Table 10-3

Capacitor	R_{eq}
C_1	
C_2	
C_3	

Figure 10-2

5. In the same manner as in step 4, you can trace the charge/discharge path for C_2 and C_3. For C_2, R_{E1} is in parallel with the capacitor and the combination of R_{E2}, r_e, and the reflected resistance of the base circuit. (The reflected resistance of the base circuit is only 4 Ω to 6 Ω because it is divided by β to move it to the emitter circuit.) Note that for C_3, the collector resistance appears to be *in series* with the load resistance. Compute the equivalent resistance seen by C_2 and C_3. Enter the computed values in Table 10-3.

6. Compute the lower critical frequency for each capacitor (C_1, C_2, and C_3) from the equation:

$$f = \frac{1}{2\pi R_{eq} C}$$

Use the R_{eq} from Table 10-3 for each capacitor. Enter the computed lower critical frequency for each capacitor in Table 10-4. The overall critical frequency of the amplifier will be higher than the highest frequency determined for each individual capacitor. To obtain a rough estimate of the upper cutoff, you can simply add the three critical frequencies; the actual frequency will be lower than this estimate. Enter the estimated overall frequency in Table 10-4.

Table 10-4

| Capacitor | -- $f_{critical}$ -- | |
	Computed	Measured
C_1		
C_2		
C_3		
Overall		

7. To measure the critical frequency due to C_1, you need to isolate this capacitor by "swamping" out the effect of C_2 and C_3. Place a 1000 µF capacitor across C_2 and another 1000 µF capacitor across C_3; this causes their frequency response to have little effect on the output. Observe the output signal in midband (around 10 kHz) and adjust the signal for 5.0 vertical divisions on the scope face. The output should appear undistorted. Reduce the generator frequency until the output falls to 70.7% (approximately 3.5 divisions) of the voltage in midband. This frequency is the lower critical frequency due to C_1. Measure and record the value in Table 10-4.

8. Using the 1000 µF capacitors, isolate C_2 by placing the large capacitors in parallel with C_1 and C_3. Measure and record the critical frequency for C_2 in Table 10-4.

9. Measure the critical frequency for C_3 by the same method. Record the value in Table 10-4.

10. Remove the large capacitors and measure the overall critical frequency of the amplifier. Record the value in Table 10-4.

Application Activity

11. Assume you need to raise the lower critical frequency to 300 Hz, as in the Application Activity. Calculate the value of a bypass capacitor, C_2, for the circuit in Figure 10-1 that will accomplish this. Record the new calculated value of C_2 in Table 10-5.

Table 10-5

Computed value of C_2	Measured low frequency

12. Modify the circuit by replacing C_2 with the closest available calculated value of C_2. Measure the new lower cutoff frequency and record it in Table 10-5.

Conclusion: Part 1

Questions: Part 1

1. For the common-source amplifier in Figure 9-1 (in the previous experiment), which capacitor (C_1, C_2, or C_3) has the most important effect on the low-frequency response? Justify your answer.

2. What change would you make to the circuit to lower the frequency response in the CE amplifier in Figure 10-1 by a factor of five?

Part 2: High-Frequency Response

1. Essentially, the circuit for this part is the same as in Part 1 (with three additional capacitors). If you did not do Part 1, measure the resistors listed in Table 10-1 and record their values before proceeding. It is also useful if you know the β_{ac} for your transistor; if you do not know β_{ac}, you can assume a typical value; for the 2N3904 a value of 200 is reasonable.

2. Calculate the ac and dc parameters listed in Table 10-6 for the CE amplifier shown in Figure 10-3. The purpose of C_4, C_5, and C_6 is to reduce the high-frequency response to make it easier to measure; they do not affect any other parameter. Record the computed values in Table 10-6.

Table 10-6

Parameter	Computed Value	Measured Value
V_B		
V_E		
I_E		
V_C		
V_{CE}		
r_e		
A_v		
V_{out}		

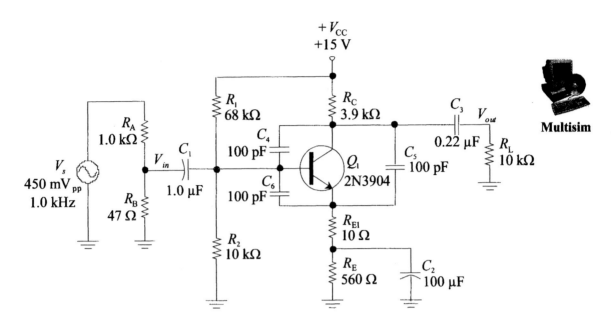

Figure 10-3

3. Construct the amplifier shown in Figure 10-3. Then measure and record the parameters listed in Table 10-6 and confirm your calculations. Recheck your work if the calculated and measured values differ significantly.

4. In this step, and in steps 5 and 6, you will compute the upper critical frequency due to the input network. Capacitors C_4, C_5, and C_6 are included in this circuit to significantly reduce the upper frequency response and make it simple to measure. If these capacitors were not present, the input capacitance, C_{in}, would be composed of just the transistor's internal base-emitter capacitance, C_{be}, and the Miller capacitance, $C_{in(Miller)}$ which is calculated from the internal base-collector capacitance. That is,
$$C_{in} = C_{be} + C_{in(Miller)}.$$

In the circuit in Figure 10-3, C_6 "swamps" C_{be}, so the base-emitter capacitance will be assumed to be just C_6 alone. Likewise, C_4 is much larger than the internal C_{bc} capacitance, so C_4 acting alone will be used to determine the input Miller capacitance. (In a small-signal transistor such as the 2N3904, the internal base-collector capacitance is typically between 3 pF and 5 pF).

By substitution,
$$C_{in} = C_6 + C_4 (|A_v| + 1) \quad \text{(use absolute value of gain)}$$

Record the input capacitance on the first line (Step 4) of Table 10-7.

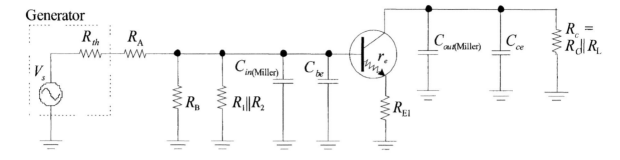

Figure 10-4

Table 10-7

Step	Parameter	Computed Value	Measured Value
4	C_{in}		
5	$R_{eq(in)}$		
6	$f_{c(in)}$		
7	C_{out}		
8	R_c		
9	$f_{c(out)}$		
10	f_{cu}		

5.   Compute the equivalent resistance, $R_{eq(in)}$, which is the discharge path for the input capacitors composed of $C_{in(Miller)}$ and C_{be}. (see Figure 10-4). $R_{eq(in)}$ is composed of four parallel paths, which are $(R_A+R_{th}) \parallel R_B \parallel R_1 \parallel R_2 \parallel (\beta_{ac}(R_{E1}+r_e))$. Enter the computed value of $R_{eq(in)}$ in Table 10-7.

6.   Compute the upper critical frequency due to the input network, $f_{c(in)}$. The frequency can be calculated using a simple RC circuit with R and C composed of the equivalent values found in steps 4 and 5. Enter the computed critical frequency in Table 10-7.

7.   In this step, and in steps 8 and 9, you will compute the upper critical frequency due to the output network. Start by finding the equivalent output capacitance, C_{out}, as illustrated in Figure 10-4. Assume C_{ce} is equal to C_5 since the added capacitor is much larger than the actual collector-emitter capacitance. This capacitance is in parallel with the output Miller capacitance. The total output capacitance, C_{out}, is found from:

$$C_{out} = C_{ce} + C_{out(Miller)}$$

$$= C_5 + C_4 \left(\frac{|A_v|+1}{|A_v|} \right) \quad \text{(use absolute value of gain)}$$

Record the output capacitance in Table 10-7.

8. Compute the equivalent resistance, R_c, seen by $C_{out(Miller)} \| C_{ce}$. The transistor looks like a current source, so the discharge path is only through $R_C \| R_L$. Enter the computed value in Table 10-7.

9. Compute the upper critical frequency due to the output network, $f_{c(out)}$. Again the circuit has been simplified to a basic RC circuit. Enter the computed value in Table 10-7.

10. The overall upper critical frequency of the amplifier will be *less* than the lowest frequency determined from the input and output networks. One way to estimate the combined effect is to use the product-over-sum rule with the two frequencies. Enter the computed overall frequency, f_{cu}, in Table 10-7. Then, observe the output signal in midband (about 1 kHz) and adjust the signal for 5.0 vertical divisions on the scope face. The output should appear undistorted. Increase the generator frequency until the output falls to 70.7% (approximately 3.5 divisions) of the voltage observed in midband. This frequency is the upper critical frequency, f_{cu}. Measure and record this frequency in Table 10-7.

Conclusion: Part 2

Questions: Part 2

1. How does the presence of the unbypassed 10 Ω resistor in the emitter circuit affect the upper critical frequency? (Hint: Consider the Miller capacitance).

2. The rise time of an amplifier, t_r, is given by the equation $t_r = 0.35/BW$.
   (a) Explain how you could measure the rise time of the amplifier to determine the upper critical frequency.

   (b) Using your measured results, calculate the rise time of the amplifier in this experiment.

Multisim Simulation

Multisim

Multisim files for the lab manual are on the website www.prenhall.com/floyd. Open the Multisim file Experiment_10_FreqResponse-nf. You can check dc parameters with a meter, ac parameters with the oscilloscope, and the frequency response with the Bode plotter. The Bode plotter is a virtual instrument that generates a graph of the frequency response. It is similar in function to a spectrum analyzer. In order to display the response of the amplifier correctly, it must be connected as shown in the Multisim file. To read the cutoff frequencies on the Bode plot, move the cursor until the signal is attenuated by 3 dB from the midband reading and note the frequency. You can do this for both the high critical frequency and the low critical frequency. The oscilloscope is not connected, but can be used to look at the signal at various points in the circuit.

Experiment 11 Thyristors

Thyristors are semiconductor devices consisting of four alternating layers of *p* and *n* material. Thyristors are primarily used in power control and switching applications. A variety of geometry and gate arrangements are available, leading to various types of thyristors such as the diac, triac, and silicon-controlled rectifier (SCR).

In Part 1, you will investigate an SCR, one of the oldest and most popular of the thyristors. It is a four-layer device that can be represented as equivalent *pnp* and *npn* transistors. In the experiment, you will compare the equivalent transistor circuit with an SCR circuit. To simulate the Application Activity, a circuit is constructed and tested using the secondary of a low voltage transformer as an ac source.

In Part 2, you will investigate a UJT oscillator circuit. The UJT has a negative resistance characteristic described by the curve in Figure 11-38 of the text. The negative resistance characteristic makes it useful for triggering circuits and oscillators as you will see.

Reading

Floyd, *Electronic Devices*, Eighth Edition, Chapter 11

Key Objectives

Part 1: Measure the gate trigger voltage and holding current for an SCR and test an SCR circuit used for an ac control application.

Part 2: Connect a UJT as a relaxation oscillator; determine the effect of varying resistance on the wave shape and the output frequency of the oscillator.

Components Needed

Part 1: The SCR

Resistors: one 160 Ω, two 1.0 kΩ, one 10 kΩ
One 10 kΩ potentiometer
One 0.1 μF capacitor
One LED
One 2N3904 *npn* transistor (or equivalent)
One 2N3906 *pnp* transistor (or equivalent)
One 2N5060 SCR (or equivalent)
One 12.6 V fused power transformer

Part 2: The Unijunction Transistor

Resistors (one of each): 47 Ω, 220 Ω, 15 kΩ
One 100 kΩ potentiometer
One 0.1 μF capacitor
One 2N2646 UJT or equivalent

Part 1: The SCR

1. Measure and record the value of the resistors listed in Table 11-1.

Table 11-1

Resistor	Listed Value	Measured Value
R_1	1.0 kΩ	
R_3	160 Ω	
R_4	1.0 kΩ	
R_5	10 kΩ	

Table 11-2

Parameter	Transistor Latch	SCR
$V_{AK(off\ state)}$		
$V_{AK(on\ state)}$		
$V_{Gate\ trigger}$		
V_{R4}		
$I_{H(min)}$		

2. Construct the transistor latch shown in Figure 11-1. The purpose of C_1 is to prevent noise from triggering the latch. The switch can be made from a piece of wire. Set R_2 for the maximum resistance and close SW1. The LED should be off. Measure the voltage across the latch shown in Table 11-2 as $V_{AK(off\ state)}$. (V_{AK} is the anode to cathode voltage.) Enter the measured voltage in Table 11-2 under Transistor Latch.

3. Slowly decrease the resistance of R_2 until the LED comes on. Measure V_{AK} with the LED on (latch closed). Record this value as $V_{AK(on\ state)}$ in Table 11-2. Measure the voltage across R_3. Record this as the gate trigger voltage in Table 11-2.

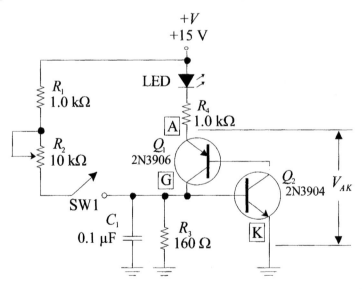

Figure 11-1

4. Open SW1. The LED should stay on because of latching action. Connect a voltmeter across R_4. Monitor the voltage while slowly decreasing the positive supply voltage. Record V_{R4} in Table 11-2. This is the smallest voltage you can obtain across R_4 with the LED on. Then apply Ohm's law to R_4 to compute the current through R_4. This current is the minimum holding current for the transistor latch. Record this as $I_{H(min)}$ in Table 11-2.

5. Replace the transistor latch with an SCR as shown in Figure 11-2. Repeat steps 2, 3, and 4 for the SCR. Enter the data in Table 11-2 under <u>SCR</u>.

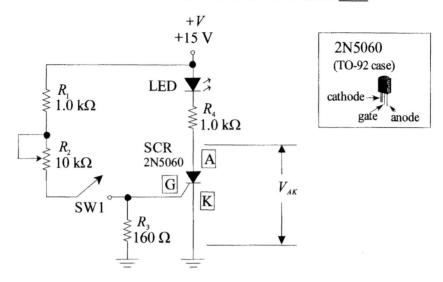

Figure 11-2

6. As you have seen, the only way to turn off an SCR is to drop the conduction to a value below the holding current. A circuit that can do this for dc operation is shown in Figure 11-3. In this circuit, the capacitor is charged to approximately +15 V (the supply voltage). When SW2 is momentarily closed, the capacitor is connected in reverse across the SCR, causing the SCR to drop out of conduction. This is called *capacitor commutation*. Modify the circuit to include this addition.

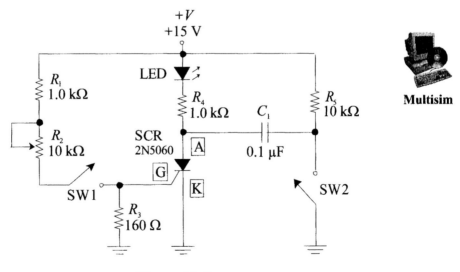

Figure 11-3

7. Test the commutation circuit by <u>momentarily</u> closing SW1 and then <u>momentarily</u> closing SW2.
Observations:

8.   For this step, be certain that you are using a properly fused and grounded transformer that has no exposed primary leads. Do not touch any connection in the circuit. Have your connections checked by your instructor before applying power to the circuit.

A typical application of SCRs is in ac circuits such as motor-speed controls. The ac voltage is rectified by the SCR and applied to a dc motor. Control is obtained by triggering the gate during the positive alteration of the ac voltage. The SCR drops out of conduction on each negative half-cycle; therefore, a commutation circuit is unnecessary. Remove the commutation circuit and replace the positive dc supply with a 12.6 V rms voltage from a low-voltage power transformer. Observe the voltage waveform across R_4 by connecting one channel of your oscilloscope on each side of R_4 and measuring the voltage difference between the probes.[1] Compare this waveform with the anode to cathode voltage across the SCR. Vary R_2 and observe the effect on the waveforms. On Plot 11-1, sketch representative waveforms across R_4 and across the SCR. Show the measured voltage on your sketch and add labels and a title.

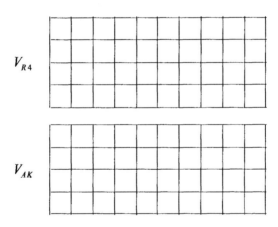

V_{R4}

V_{AK}

Plot 11-1

Conclusion: Part 1

---

[1] To find the voltage difference, set both channels to the same VOLTS/DIV setting. Most oscilloscopes will have an ADD control. ADD the channels and INVERT CH 2.

Questions: Part 1

1. For the circuit of Figure 11-3, what effect on the voltage waveform measured across R_4 would you expect if the holding current for the SCR were higher?

2. How does capacitor commutation in the circuit in Figure 11-3 work?

Part 2: The Unijunction Transistor

1. Measure and record the resistance of the resistors listed in Table 11-3.

Table 11-3

Resistor	Listed Value	Measured Value
R_1	47 Ω	
R_2	220 Ω	
R_3	15 kΩ	

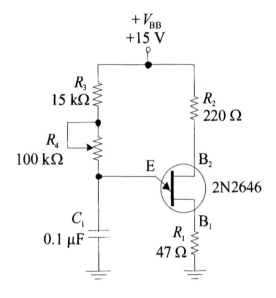

Figure 11-4

2. Construct the circuit shown in Figure 11-4. Set R_4 for zero resistance. Observe the waveforms at the emitter, base 1, and base 2. Sketch the observed waveforms in the proper time relationship on Plot 11-2. Label the scales with the time and amplitude.

103

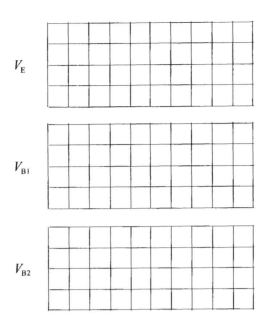

V_E

V_{B1}

V_{B2}

Plot 11-2

3. Set the potentiometer for each resistance listed in Table 11-4. You will need to remove power from the circuit and disconnect one end of the potentiometer to do this. Add the measured value of R_3 to determine the total resistance at each setting. Then measure the frequency at each resistance setting and record the total resistance and frequency data in Table 11-4.

Table 11-4

Resistance setting of R_4	Total Resistance	Measured Frequency
5.0 kΩ		
25 kΩ		
45 kΩ		
65 kΩ		
85 kΩ		

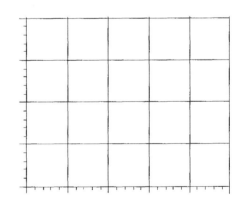

Resistance (kΩ)
Plot 11-3

4. Graph the frequency as a function of the total resistance in Plot 11-3. What conclusion can you draw from this data?

5. Describe the effect on the emitter waveform as the potentiometer is varied.

_____

6. Try reducing the power supply voltage while you observe the waveform at the emitter. Does this affect the frequency? The amplitude?

_____

Conclusion: Part 2

Questions: Part 2
1. Assume a UJT has an intrinsic standoff ratio of 0.6 and V_{BB} is 10 V. What value of V_E will just turn on the UJT?

2. What conditions must be satisfied before the UJT can turn on?

Multisim Simulation

Multisim files for the lab manual are on the website www.prenhall.com/floyd. Open the Multisim file Experiment_11_SCR-nf. The Multisim circuit simulates the SCR circuit in Figure 11-3. Determine the firing point for the SCR. Compare the simulation with your experimental result.

Multisim

Experiment 12-A Operational Amplifiers

The 741 operational amplifier ("op-amp") was designed in 1968 and is still popular because it can meet requirements for many basic applications. Although newer designs have much improved specifications, the 741 is a reasonable choice for many circuits. You will use a 741C in this experiment and subsequent experiments.

Most op-amps have a differential amplifier as the input stage. There are important advantages to this type of input amplifier, particularly for noise rejection. In Part 1, you will investigate a differential amplifier made from discrete components. Details of this amplifier are discussed in Chapter 6 of the text, but covered here with operational amplifiers.

In Part 2, several important specifications for op-amps are described and measured. These include input offset voltage, input bias current, input offset current, CMRR', and slew rate.

In Part 3, inverting and noninverting amplifiers are constructed. The inverting amplifier is investigated first; this circuit is the starting point for many other op-amp circuits, including summing amplifiers and integrators. The non-inverting amplifier is equivalent to the one in the Application Activity of the text.

Following part 3 is an optional software exercise as a Programmable Analog Design feature using the AnadigmDesigner2 program. The program is free (at www.anadigm.com) and easy to learn. The exercise gives step-by-step instructions for creating a working analog circuit that could be downloaded into a programmable analog array as a working circuit.

Reading

Floyd, *Electronic Devices*, Eighth Edition, Chapter 12 and review Section 6-7, The Differential Amplifier

Key Objectives

Part 1: Construct and test a discrete differential amplifier with current-source biasing.

Part 2: Measure the input offset voltage, input bias current, input offset current, CMRR', and slew rate for a 741C op-amp.

Part 3: Construct and test inverting and noninverting amplifiers using op-amps.

Components Needed

Part 1: The Differential Amplifier

Resistors: two 100 Ω, one 4.7 kΩ, three 10 kΩ, one 33 kΩ, two 100 kΩ
Capacitors: two 10 μF
Transistors: three 2N3904

Part 2: Op-Amp Specifications

Resistors: two 100 Ω, two 10 kΩ, two 100 kΩ, one 1.0 MΩ
Two 1.0 µF capacitors
One LM741C op-amp

Part 3: Basic Op-Amp Circuits

Resistors: two 1.0 kΩ, one 10 kΩ, one 150 kΩ, one 470 kΩ, one 1.0 MΩ
Two 1.0 µF capacitors
One LM741C op-amp

Part 1: The Differential Amplifier

1. Measure and record the values of the resistors listed in Table 12-1. Best results can be obtained if R_{B1} and R_{B2} are matched and R_{E1} and R_{E2} are also matched.

Table 12-1

Resistor	Listed Value	Measured Value
R_{B1}	100 kΩ	
R_{B2}	100 kΩ	
R_{E1}	100 Ω	
R_{E2}	100 Ω	
R_T	10 kΩ	
R_{C2}	10 kΩ	

Table 12-2

DC Parameter	Computed Value	Measured Value
V_A	−1 V	
I_T		
$I_{E1} = I_{E2}$		
$V_{C(Q1)}$		
$V_{C(Q2)}$		

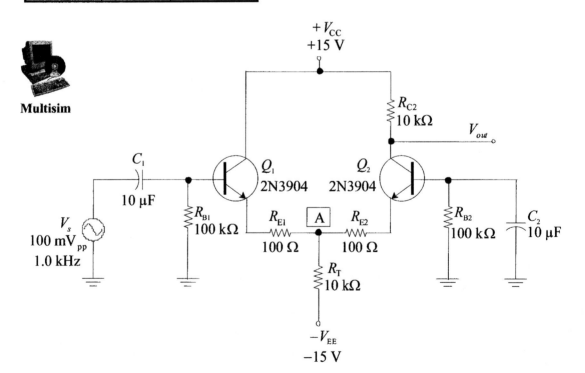

Figure 12-1

108

2. Construct the single–ended differential amplifier shown in Figure 12-1. Compute and record the dc parameters listed in Table 12-2. The transistor bases are slightly above ground due to the small base current. This causes the voltage at point A to be about -1 V (because of the 0.7 V base-emitter drop). This assumption simplifies finding the current in R_T, which is sometimes referred to as the "tail" resistor. The tail current, I_T, can be found by applying Ohm's law to R_T. (An exact solution can be found by applying Kirchhoff's voltage law around one of the base-emitter paths).

3. Measure and record the dc parameters listed in Table 12-2.

4. Compute and record the ac parameters given in Table 12-3 except $A_{v(cm)}$ and CMRR', which are found in step 6 and step 8. The differential amplifier can be thought of as a common-collector amplifier (Q_1) driving a common-base amplifier (Q_2). For the single-ended input signal, the differential voltage gain is given by:

$$A_{v(d)} = \frac{R_{C2}}{2\left(R_{E2} + r_e\right)}$$

Except for the 2 in the denominator, this equation is equivalent to the gain equation for a CB amplifier. The reduction by a factor of 2 is due to the attenuation of the signal to point A by the CC amplifier. To compute $R_{in(tot)}$, begin by assuming $r_{e(Q2)}$ is in series with R_{E2}, R_{E1}, and $r_{e(Q1)}$. Move this resistance into the base circuit of Q_1 by multiplying by β_{Q1}. This result is then seen to be in parallel with the Q_1 base resistor, R_{B1}. Writing this in equation form:

$$R_{in(tot)} = R_{B1} \,||\, \left\{\beta_{Q1}\left(r_{e(Q1)} + R_{E1} + R_{E2} + r_{e(Q2)}\right)\right\}$$

If you don't know β_{Q1}, assume a typical value of 200 for the 2N3904.

Table 12-3

AC Parameter	Computed Value	Measured Value
$V_{b(Q1)}$	100 mV$_{pp}$	
V_A		
$r_{e(Q1)} = r_{e(Q2)}$		
$A_{v(d)}$		
$V_{c(Q2)}$		
$R_{in(tot)}$		
$A_{v(cm)}$		
CMRR'		

5. Add the single-ended differential mode ac signal as shown in Figure 12-1 and measure the ac parameters listed in Table 12-3 (except $A_{v(cm)}$ and CMRR'). To determine the measured value of $R_{in(tot)}$, first note the output voltage, V_c; then add a 33 kΩ test resistor, R_{test}, in series with the input signal. The test resistor forms a

voltage divider with the input resistance and will cause the output voltage to drop. Call the reduced output voltage, V_c'. The value of $R_{in(tot)}$ is determined by solving for $R_{in(tot)}$ from the ratio:

$$\frac{V_c'}{R_{in(tot)}} = \frac{V_c}{R_{test} + R_{in(tot)}}$$

Solving for $R_{in(tot)}$:

$$R_{in(tot)} = \left(\frac{V_c'}{V_c - V_c'}\right) R_{test}$$

Notice that if the output is halved after inserting the test resistor, $R_{test} = R_{in(tot)}$.

6. Calculate the common-mode gain, $A_{v(cm)}$. The common-mode gain can be approximated from the formula:

$$A_{v(cm)} \cong \frac{R_C}{2R_T}$$

This formula is based on the assumption that the two sides of the differential amplifier are balanced. Enter the computed common-mode gain in Table 12-3.

7. The common-mode gain is the gain observed when the same signal is applied to both sides of the differential amplifier. Remove the test resistor from step 5 and connect the circuit shown in Figure 12-2, in which the input signal is applied in common-mode. In order to measure the common-mode gain, raise the input signal level from the signal generator until a 1 V_{pp} output is observed. Then measure the ratio of the output to input signal and record the measured $A_{v(cm)}$ in Table 12-3.

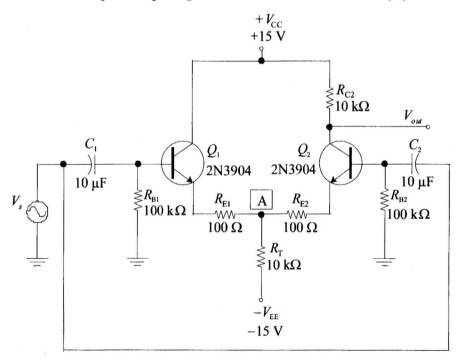

Figure 12-2 Applying a common-mode signal to the differential amplifier.

110

8.	CMRR′ is the decibel common-mode rejection ratio (indicated with the prime symbol). It is 20 times the logarithmic ratio of the absolute value of the ratio of $A_{v(d)}$ to $A_{v(cm)}$, expressed in dB. In equation form, this is

$$CMRR' = 20\log\left|\frac{A_{v(d)}}{A_{v(cm)}}\right|$$

Use the calculated value of $A_{v(d)}$ and $A_{v(cm)}$ to obtain the calculated value of CMRR′. Then perform the same calculation but using the measured gain of $A_{v(d)}$ and $A_{v(cm)}$ to obtain the measured value of CMRR′. Record both the calculated and measured values in Table 12-3.

Adding a Constant-Current Source

9.	A constant-current source can improve the ability to reject common mode signals by reducing the common-mode gain. Replace the tail resistor with the constant-current source consisting of three resistors and a BJT as shown in Figure 12-3. Check that the three added resistors are close to their specified values. Then measure the common-mode gain with the constant-current source. Observations:

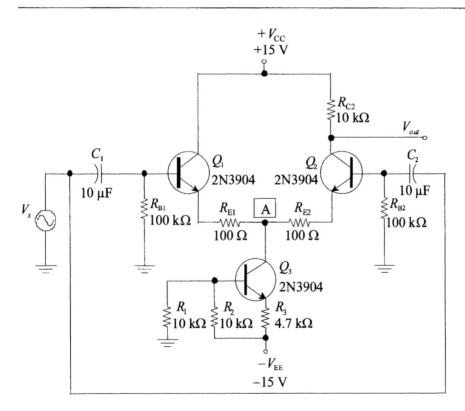

Figure 12-3

Conclusion: Part 1

Questions: Part 1

1. In step 5, you were directed to measure the input resistance while observing the output voltage, V_c, instead of the input voltage, V_b. What advantage does this method have to assure a good measurement?

2. How much current is sourced by Q_3 to the differential amplifier in Figure 12-3?

Part 2: Op-Amp Specifications

The 741C is available as a DIP (dual-inline pins) package as shown. Pin 1 is usually indicated with a dot or is on the end of the IC with a notch. Pin numbers are assigned counterclockwise, starting from the top left corner as shown in Figure 12-4.

pin 1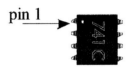

Figure 12-4

1. Table 12-4 lists important specifications for the 741C op-amp for $T_A = 25°C$. You may want to confirm these specifications on the manufacturer's specification sheet, which is available at http://www.fairchildsemi.com/ds/KA/KA741.pdf.

Table 12-4

Step	Parameter	Specified Value			Measured Value
		Minimum	Typical	Maximum	
2d	Input offset voltage, V_{OS}		2.0 mV	6.0 mV	
3d	Input bias current, I_{BIAS}		80 nA	500 nA	
3e	Input offset current, I_{OS}		20 nA	200 nA	
4b	Differential gain, $A_{v(d)}$				
4c	Common-mode gain, $A_{v(cm)}$				
4d	CMRR′	70 dB	90 dB		
5	Slew rate		0.5 V/µs		

2. Measure the input offset voltage, V_{OS}, of a 741C op-amp. The input offset voltage is the amount of voltage that must be applied between the input terminals through two equal resistors to give zero output voltage. It is a dc parameter.

(a) Measure and record the values of the resistors listed in Table 12-5. R_C is for bias compensation.

(b) Connect the circuit shown in Figure 12-5.

(c) Measure the output voltage, V_{OUT}. The input offset voltage is found by dividing the output voltage by the closed-loop gain. You should use a noninverting amplifier configuration and your measured resistors for the purpose of the offset calculation (see text Equation 12-8).

(d) Record the measured V_{OS} in the first line in Table 12-4.

Table 12-5

Resistor	Listed Value	Measured Value
R_f	1.0 MΩ	
R_i	10 kΩ	
R_C	10 kΩ	

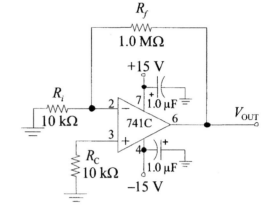

Figure 12-5 V_{OS} measurement.

3. In this step, you will measure the input bias current, I_{BIAS}, and the input offset current, I_{OS}, of a 741C op-amp. The input bias current is the average of the input currents at each input terminal. The input offset current is a measure of how well these two currents match. The input offset current is the difference in the two bias currents when the output voltage is 0 V. The input bias current and input offset current are dc parameters.

(a) Measure and record the values of the resistors listed in Table 12-6.

(b) Connect the circuit shown in Figure 12-6.

(c) Measure the voltage across R_1 and R_2 of Figure 12-6. Use Ohm's law to calculate the current in each resistor.

(d) Record the *average* of these two currents in Table 12-4 as the input bias current, I_{BIAS}.

(e) Record the *difference* in these two currents in Table 12-4 as the input offset current, I_{OS}.

Table 12-6

Resistor	Listed Value	Measured Value
R_1	100 kΩ	
R_2	100 kΩ	

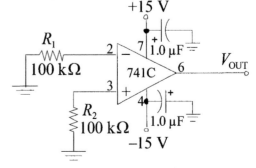

Figure 12-6 I_{BIAS} and I_{OS} measurement.

113

4. Measure the common-mode rejection ratio, CMRR', of a 741C op-amp. The CMRR' is 20 times the logarithmic ratio of the op-amp's differential gain ($A_{v(d)}$) divided by the common-mode gain ($A_{v(cm)}$). Because it is a ratio of gains, CMRR' is an ac parameter. It is frequently expressed in decibels (and shown with the prime symbol to indicate this). The definition for CMRR' is

$$\text{CMRR}' = 20 \; \log \left| \frac{A_{v(d)}}{A_{v(cm)}} \right|$$

(a) Measure and record the values of the resistors listed in Table 12-7. For an accurate measurement of CMRR', resistors R_A and R_B should be closely matched as should R_C and R_D.

(b) It is more accurate to compute the differential gain, $A_{v(d)}$, based on the resistance ratio than to measure it directly. Determine the differential gain by dividing the measured value of R_C by R_A. Enter the differential gain, $A_{v(d)}$, in Table 12-4.

(c) Connect the circuit shown in Figure 12-7. Set the signal generator for 1.0 V_{pp} at 1.0 kHz. Measure the output voltage, $V_{out(cm)}$. Determine the common-mode gain, $A_{v(cm)}$, by dividing $V_{out(cm)}$ by $V_{in(cm)}$. Record the result in Table 12-4.

(d) Determine the measured value of CMRR' (in decibels) for your 741C op-amp. Record the result in Table 12-4.

Table 12-7

Resistor	Listed Value	Measured Value
R_A	100 Ω	
R_B	100 Ω	
R_C	100 kΩ	
R_D	100 kΩ	

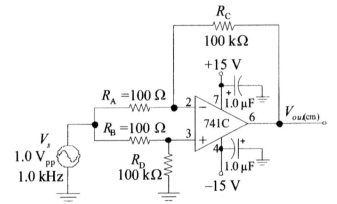

Figure 12-7 CMRR' measurement.

5. Measure the slew rate of your op-amp. Slew rate is the internally limited rate of change in output voltage with a large-amplitude step function applied to the input. It affects large signals more than small signals. It is usually specified for a unity-gain voltage-follower with a fast rising input pulse. Slew rate is usually expressed in units of volts/microsecond (V/μs).

Connect the unity gain circuit shown in Figure 12-8. Set the signal generator for a 10 V_{pp} square wave at 10 kHz. The output voltage will be slew-rate limited and will not respond instantaneously to the change in the input voltage. The slew rate can be measured by observing the change in voltage divided by the change in time at any two points on the rising output waveform as shown in Figure 12-9. Record the measured value in Table 12-4.

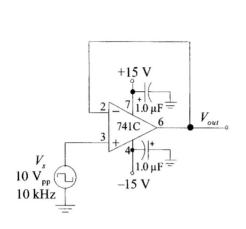

Figure 12-8 Slew-rate measurement.

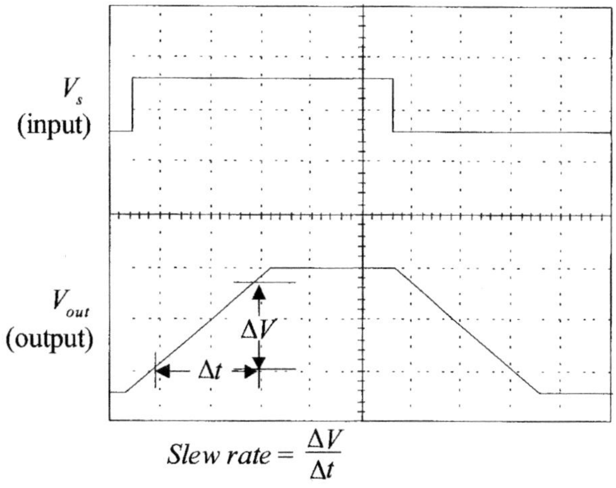

Figure 12-9

$$Slew\ rate = \frac{\Delta V}{\Delta t}$$

Conclusion: Part 2

Questions: Part 2
1. What is the difference between the input bias current and the input offset current?

2. What is the advantage of a high value of CMRR'?

Part 3: Basic Op-Amp Circuits
Inverting Amplifier
1. You will use a 741C op-amp for constructing the circuits in this part. The pin diagram was given in Part 2 as Figure 12-4.
    (a) The schematic for the inverting amplifier is shown in Figure 12-10. Notice that 1.0 µF bypass capacitors are shown on the power supply leads; these are added to avoid noise problems when working with op-amps. Note the polarities of the capacitors. Measure a 1.0 kΩ resistor for R_1 and a 10 kΩ resistor for R_f. (R_1 is the input resistor; R_f is the feedback resistor). Record the measured resistors in Table 12-8.

115

(b) Using the measured resistance, compute the closed-loop gain of the inverting amplifier. Record the computed value in Table 12-9.

(c) Calculate V_{out} using the computed closed-loop gain and record it as the computed value in Table 12-9.

(d) Connect the circuit shown in Figure 12-10. Set the input for a 500 mV$_{pp}$ sine wave at 1.0 kHz with no dc offset. Measure and record V_{in} in Table 12-9.

(e) Measure and record the output voltage, V_{out}.

(f) Measure and record the voltage at pin 2. This point is called a *virtual ground*. A virtual ground is not actually at ground potential, (or else there would be nothing to amplify!), but the voltage is near ground by the effect of feedback.

(g) Place a 1.0 kΩ test resistor (R_{test}) in series with the generator and R_i. Measure the input resistance of the circuit, R_{in}, by observing the output voltage. In this case, you should observe that the output is halved, indicating that the resistance of the test resistor and the input resistance of the amplifier are about the same. Record the measured value of R_{in} in Table 12-9.

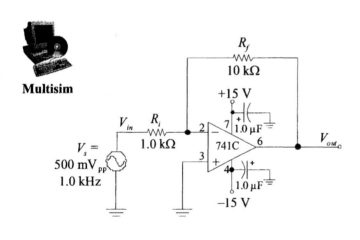

Multisim

Figure 12-10

Table 12-8

Resistor	Listed Value	Measured Value
R_1	1.0 kΩ	
R_f	10 kΩ	

Table 12-9

Parameter	Computed Value	Measured Value
V_{in}	500 mV$_{pp}$	
$A_{cl(1)}$		
V_{out}		
$V_{(-)}$		
R_{in}		

Noninverting Amplifier

2. The circuit for this step is the noninverting amplifier shown in Figure 12-11.

(a) Using the measured resistances from step 1, compute the closed-loop gain of the noninverting amplifier. The closed-loop gain equation is given in the text as Equation 12-8. Enter the computed value in Table 12-10.

(b) Calculate V_{out} using the computed closed-loop gain. Record in Table 12-10.

(c) Change the circuit to the noninverting amplifier circuit shown in Figure 12-11. Set the input for a 500 mV$_{pp}$ sine wave at 1.0 kHz with no dc offset. Record the measured setting in Table 12-10.

(d) Measure the output voltage, V_{out}. Record the measured value in Table 12-10.

(e) Measure the feedback voltage at pin 2. Record the measured value. Notice that the voltage at pin 2 is not near ground potential this time.

(f) Place a 1.0 MΩ test resistor in series with the input from the generator. Observe the output voltage with the series resistor in place. You can think of the voltage change as being dropped across the test resistor, the rest is across the input resistance. You can use this to indirectly find the input resistance. Record the measured input resistance in Table 12-10.

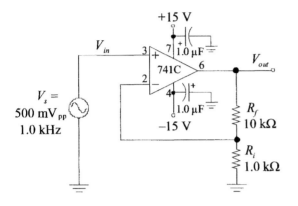

Table 12-10

Parameter	Computed Value	Measured Value
V_{in}	500 mV$_{pp}$	
$A_{cl(NI)}$		
V_{out}		
$V_{(-)}$		
R_{in}		

Application Activity

Figure 12-11

3. Change R_f to 150 kΩ and reduce the input signal to 50 mV$_{rms}$ (141 mV$_{pp}$) as shown in Figure 12-12. (This is equivalent to the Application Activity op-amp circuit at maximum gain.) Check the gain at 1.0 kHz. Then, raise the frequency to 10 kHz. Describe your observations. The gain change is due to the higher frequency. The effect of higher frequencies on the closed-loop gain is discussed in the text in Section 12-8.

Observations:

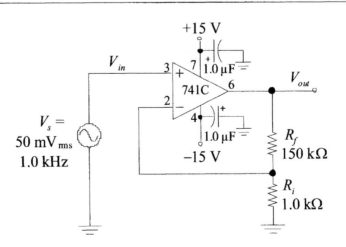

Figure 12-12

Conclusion: Part 3

Questions: Part 3

1.   Express the gain of the two amplifiers tested in steps 1 and 2 in dB.

2.   It was correct to talk about a *virtual ground* for the inverting amplifier. Why isn't it correct to refer to a virtual ground for the noninverting amplifier?

3.   Why was the gain lower in step 3 when the frequency was increased?

Multisim Simulation

Multisim

Multisim files for the lab manual are on the website www.prenhall.com/floyd. There are three Multisim files for this experiment. The first is named Experiment_12_DiffAmp-nf and is a simulation of the circuit in Figure 12-1. You can check the dc and ac parameters against your experimental result.

Experiment_12_Inverting_Amp-nf is the second file. It is similar to the inverting amplifier in Figure 12-10 of Part 3, except for the feedback resistors, which control the gain. Determine what the gain should be, then measure it. Then determine the frequency response of the circuit. You can get a pretty good idea by using the Bode plotter, which you will need to set up. Then set the input frequency at the expected cutoff and see if the output drops to 70.7% of the value at 1.0 kHz (midband). This is a common way to measure the upper cutoff frequency.

The third file is the same inverting amplifier but with a fault. The file name is Experiment_12_Inverting_Amp-fault. Troubleshoot it and locate the fault.

Experiment 12-B Programmable Analog Design

Field Programmable Gate Array (FPGA) devices were developed in the 1980s so that logic circuits could be reconfigured without a hardware change. The Field Programmable *Analog* Array (FPAA) is a new extension of this idea that has been developed by several different semiconductor manufacturers. This experiment will give you experience with an FPAA from the Anadigm Company. Anadigm refers to its product as a *dynamically programmable* Analog Signal Processor or *dp*ASP. For this manual we will simplify this to Analog Signal Processor or ASP.

In Part 1, you will learn about the Anadigm-supplied computer-aided design (CAD) software, which is called AnadigmDesigner2 (AD2), a free download from Anadigm. AnadigmDesigner2 is used for configuring the ASP and you will practice using its simulation tool to see the signal waveforms on the PC screen. Part 1 is entirely a software exercise to load and begin using the AnadigmDesigner2 CAD software from the Anadigm website (www.anadigm.com). AnadigmDesigner2 (AD2) is an excellent CAD tool and easy to learn. The exercise in Part 1 gives step-by-step instructions for creating a configuration file that is ready to be downloaded into an ASP to become a working circuit. In addition, the Quick Start Guide that is supplied with the Programmable Analog Module (PAM) is included in Appendix B of this manual and will guide you step-by-step through the process.

In Part 2, you will transmit the configuration file created in Part 1 to an actual ASP and compare its operation to the simulation. The ASP is the heart of the Programmable Analog Module supplied by the Servenger Company (www.servenger.com). This circuit board provides convenient access to the Anadigm ASP together with on-board power supplies, input and output buffers, and electrical connectors. After configuring the circuit, you will test its operation and change the frequency and gain. As an option, you can plug self-powered stereo speakers directly into the PAM and hear the effect of these changes to the circuit. This is an excellent way to see how the ASP can be easily changed with a different circuit.

Reading

Floyd, *Electronic Devices*, Eighth Edition, Chapter 12, with emphasis on the Programmable Analog Design section

Key Objectives

Part 1: Create a simple analog circuit design using the AnalogDesigner2 CAD software and observe a simulation of the circuit.

Part 2: Download the configuration file into an Anadigm ASP on a PAM board and measure the output signals with your oscilloscope. Compare the oscilloscope displayed waveforms to the simulation waveforms.

Components Needed

Part 1: Introduction to AnadigmDesigner2

Computer with a serial port with the AnadigmDesigner2 (AD2) software installed.

Part 2: Downloading the Configuration File

Programmable Analog Module (Servenger PAM unit that includes the Anadigm Analog Signal Processor)
Serial cable (supplied with the PAM)
Power source (supplied with the PAM)
Set of self-powered stereo PC speakers, each with a 3.5 mm stereo plug

Part 1: Introduction to AnadigmDesigner2

1. If you have not yet downloaded and installed the AnadigmDesigner2 (AD2) CAD software from the Anadigm website (www.anadigm.com), refer to the PAM Quick Start Guide in Appendix B for step-by-step instructions for the download. Then open AD2. On the welcome screen, choose *Continue with default chip*.

2. You will see a blank design window like the one shown in Figure 12-13. The rectangles on the left represent the Input/Output (I/O) cells. The rectangles on the right represent the two dedicated output cells. Notice that one of the I/O cells has four inputs that connect through a 4-to-1 signal multiplexer to a single connection point (the little red box on your screen next to the label InputCell4).

    All of the I/O cells can be configured as either an input cell or an additional output cell. The default state is as an input cell.

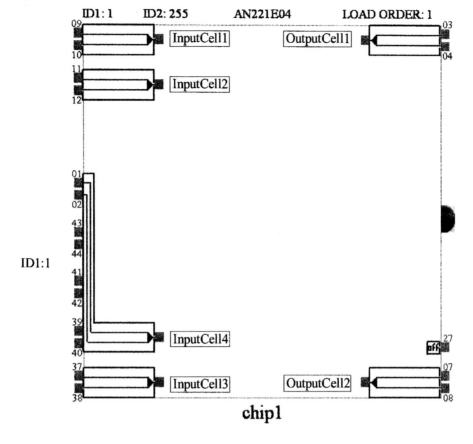

Figure 12-13 Blank design window. Labels have been added to the cells.

3.   Double-click in InputCell1, the top rectangle on the upper left side of the design window. This will open a parameter-setting window for this I/O cell. Under Options, select *Anti-Alias Filter* as OFF (default) and then select *Input* as Differential (default). The shape of the artwork symbol in the symbol box in the upper right should be as shown in Figure 12-14. Notice the explanatory text in the window area to the left of the symbol box. This text will change as you select different parameter settings. For now, accept the default settings for all the I/O cells and for the output cells. Click OK to accept these parameter settings for this cell and close the window.

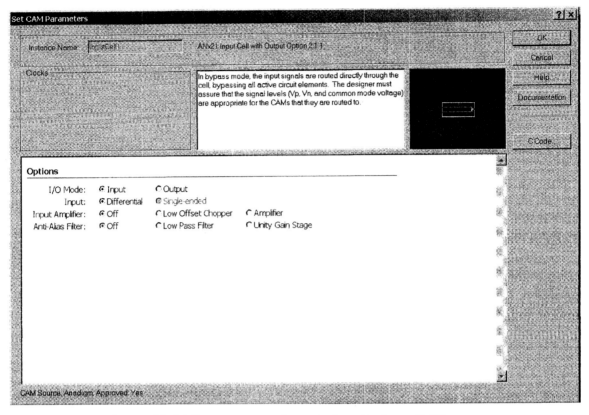

Figure 12-14 Parameter-setting window for InputCell1.

4. From the Edit menu, select *Insert new CAM*. This will bring up the library of pre-defined analog functions called configurable analog modules (CAMs) as shown in Figure 12-15. You will select CAMs from this list and then connect them together to create the circuit design.

    To create a sine wave oscillator, select *OscillatorSine* from the list and then click *Create CAM*. A ghosted image of the CAM icon will appear attached to the end of the cursor. Position the CAM inside the design window and click the left mouse button one time to drop it there.

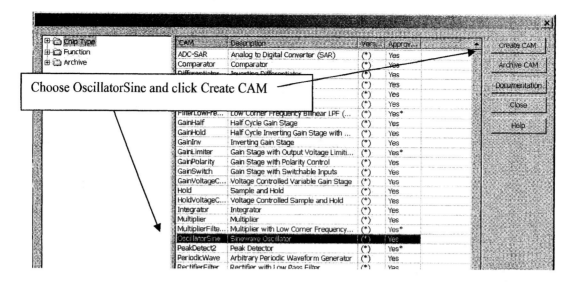

Figure 12-15 CAM selection dialog window.

5. Immediately after placing the CAM, a new parameter-setting window will appear for the sine wave oscillator CAM you have just placed. Enter a new peak amplitude of 3.0 (V) and press Enter. Notice the short message that tells you how close to this value the ASP will realize (3.02 V in this case). This is the standard amplitude measurement of the sine wave for AnadigmDesigner2; the peak-to-peak value would be twice this peak amplitude.

    Enter a new frequency of 50.0 (kHz) and then press Enter. Notice the message that 50.0 kHz will be realized and the available frequencies are 48.0 kHz to 800 kHz. Click OK to close the parameter-setting window to accept these values. You can always go back to this window to change a parameter by double-clicking on the CAM icon in the design window.

6. In this step, you will add another CAM. Press the M key to bring up the *Insert New CAM* window or click the *Get New CAM* icon (same meaning) on the menu bar – look for the green amplifier symbol with the star burst above it:

    This brings up the library of predefined CAMs as before. Select an Inverting Gain Stage *GainInv* and place it in the design window below and to the right of the sine wave oscillator.

7. When the parameter-setting window appears, set the gain to 0.333 and then close the window. Because this is an inverting amplifier this means a gain of –0.333. This is actually an attenuation, but within the ASP, there are advantages to having this stage. In Part 2, you can adjust this value to one with more gain.

    Your design window should look like Figure 12-16. If you want to move an existing CAM icon, just drag it to the new location.

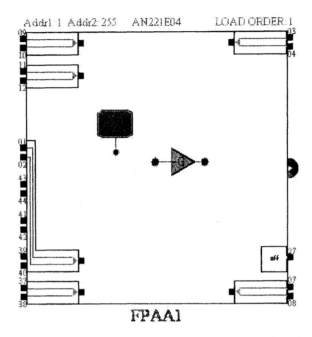

Figure 12-16 Design window with 2 CAMs in place.

8.   You will now connect the electrical path between the CAMs and an output cell. Note that connection points for each CAM are shown as small solid red dots on the screen.

Point the cursor to the connection point below the sine wave oscillator. The wiring tool will automatically appear (it looks like a thick white pen). Now click once and draw a simulated wire to the input connection point of the amplifier and then click one more time to make the connection. Now connect the output of the amplifier to OutputCell1 as shown in Figure 12-17.

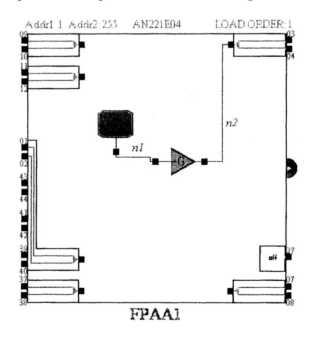

Figure 12-17 Wiring for the CAMs.

9. In this step, you will connect four simulated oscilloscope probes to display and measure the operation of the circuit. For the purpose of the simulation, the probes can be connected inside the ASP even though this is not actually possible with a real device.

    The sine wave oscillator creates an oscillating waveform with a peak value of 3 V that is amplified (by –1/3) and delivered to the output of the ASP device. You will see these waveforms on the built-in oscilloscope.

    From the Simulate menu, choose *Create Oscilloscope Probe* (or press the P key or click the scope probe icon on the tool bar). As you place each scope probe, notice that they are colored (violet, green, blue, and yellow in order). You will see the corresponding color for the traces on the oscilloscope display.

    Place the first scope probe on the output of the oscillator. Click the icon again and place another scope probe on the output of the amplifier. As shown in Figure 12-18 place a third probe on the upper outboard pin of the OutputCell1 (labeled as pin 3 and called P for Positive). Finally, place the fourth probe on the lower outboard pin of OutputCell1 (labeled as pin 4 and called N for Negative).

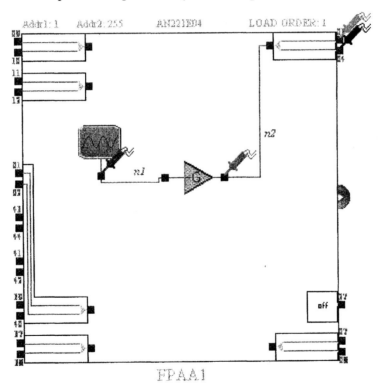

Figure 12-18 Circuit with scope probes attached for simulation.

10. Before simulating the circuit, choose *Setup Simulation* from the Simulate menu. Change the end time for the simulation to 200 µs (shown in AD2 as 200 u). This is illustrated in Figure 12-19. Accept the change by clicking OK.

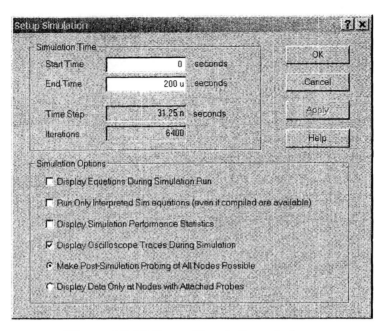

Figure 12-19 Setup Simulation window.

11. Observe the simulated circuit in operation by choosing *Begin Simulation* in the Simulate menu (or select the *Sim* icon or press the F5 key). An oscilloscope image will appear on the screen and the waveform traces will appear. Circuit simulators are computationally intensive so you may need to wait a few seconds for the simulation to run to completion (the word *Sim.* appears in place of the cursor arrow while the simulation is running). The time record is wider than the oscilloscope display which is typical for digital oscilloscopes. Use the slider below the scope display to move through the full record. Point to the Time Per Division indicator to the right of the oscilloscope display and adjust the value to 5 us. The Volts Per Division settings should all be left on the default value of 1.0 V/div.

You should see these traces on the simulation:

Violet: Output of the sine wave oscillator from 0 V to +3 V to 0 V to –3 V to 0 V repeatedly.

Green: Output of the inverting amplifier from 0 V to –1 V to 0 V to +1 V to 0 V repeatedly (same as oscillator output but with –1/3 amplitude).

(The violet and green traces are inside the ASP device. These signals are shown as *Ground Referenced* which means centered about ground or 0 V.)

Blue: A sine wave external to the ASP that looks like the output of the amplifier but with amplitude that is ± 1/2 V around a +2 V reference level.

Yellow: A sine wave that looks like the blue trace but with opposite phase.

The ASP delivers external signals as *differential signals* centered around +2 V. The individual output signals are ½ the 1 V amplitude out of the internal amplifier, but taken differentially (OutputCell1P minus Outputcell1N) the output signal amplitude is actually 1 V. This is the same as the output of the amplifier.

Sketch the results of your observation in Plot 12-1. Add the V/div and time/div labels to the plot.

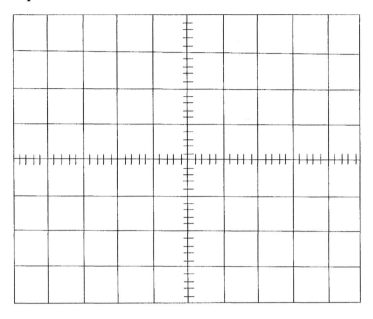

Plot 12-1 Waveforms for 3 V 50 kHz sine wave amplified by −1/3.

12.   Summarize your observations from Step 11.

_____

_____

_____

_____

13.   Save the circuit design configuration file as Exp12B (*File > Save As*, etc.) Notice what the extension is for this file type and make a note of the file name and path:

_____

Conclusion: Part 1

Questions: Part 1

1. (a) If the gain of an amplifier is specified as 1/4 (0.250), what is the output voltage for a steady-state input of +1 V?

    (b) If the gain of an amplifier is specified as –1/3 (–0.333), what is the output voltage for a steady-state input of +1 V?

2. What does it mean for a sine wave signal to be ground referenced?

3. Does a 1 V peak amplitude sine wave referenced to +2 V ever achieve a voltage amplitude of 0 V? Explain.

4. (a) What is the mathematical definition of a differential signal from a dual output signal source that has its output ports labeled as OutputA and OutputB?

    (b) If voltage waveforms appearing at OutputA and OutputB are both referenced to +2 V, would the differential signal be the same if both were referenced to 0 V?

5. What is the file name extension for Anadigm ASP configuration files saved out of AnadigmDesigner2?

Part 2: Downloading the Configuration File

For this part, and all remaining B experiments, you will need to have the PAM connected to the PC and AnadigmDesigner2 running. If you have not yet connected the PAM board to the pc, see the PAM Quick Start Guide in Appendix B for detailed instructions. Please pay careful attention to the information about USB-to-Serial adapters which are not recommended.

1. Using AD2 select *File > Open* and select the configuration file that you saved from step 13 of Part 1 with the filename Exp12B.ad2. You should see the circuit design that you previously created. If the Resource Panel shows up to the right of the design window, you can click the little corner X to close it, as it is not needed at this point.

2. Send the configuration file to the PAM with one of the following options:
    (a) *Configure > Write Configuration Data to Serial Port*, or
    (b) The Ctrl-W key combination, or
    (c) The Download icon with the blue downward pointing arrow.

    A successful configuration download is indicated by two additional green LEDs on the PAM turning ON and staying ON. The new lights mean:
    (a) Green LED near the Serial Port connector – this is the microprocessor *Successful Command Execution* indicator (the red LED would indicate an unsuccessful execution)
    (b) Green LED near the 8-pin DIP sockets – this is the ASP *Configuration is Executing* indicator.

3. Use your oscilloscope to measure the signal at the terminal post labeled OUT1 (ground the scope lead on the terminal post labeled GND). Observe the mid-level of the sine wave, the peak-to-peak voltage, and the period.
    Sketch the results of your observation in Plot 12-2. Add the V/div and time/div labels to the plot.

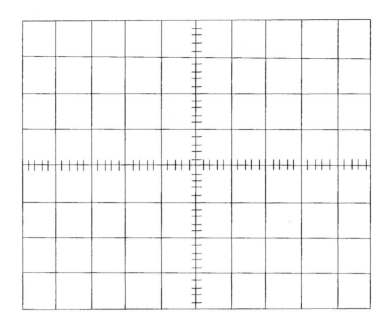

Plot 12-2 Measured waveform.

Looking back at Plot 12-1, what waveform on that previous plot does this resemble?

_____

4. Go back to the AD2 design window. There should be a green probe at the output of the Inverting Gain Stage, which corresponds to the green trace on the simulation. Run the simulation again, set the Time Per Division to 5 μs, and compare your sketch in Plot 12-2 to the waveforms on the simulated oscilloscope. All three simulated waveforms should show the same mid-level dc voltage and period, but you should see that the amplitude is slightly less from the PAM than in the simulation.

    Observations:

    _____

    _____

5. The reason for the slightly lower amplitude is due to a filter on the PAM that is not part of the ASP output. (Filters are studies later in Chapter 15.) Recall that you initially set the gain of the inverting amplifier to 0.333. (See Part 1, step 7.) Calculate a new gain for the gain stage that will produce an output from the PAM with 1.0 V_p amplitude (2.0 V_{pp}).

    Required gain for 1.0 V_p from PAM = _____

6. Using AD2, double-click on the inverting amplifier and set the gain to the value calculated in step 5. Download the new configuration file and measure the output amplitude.

    Measured output amplitude: _____

7. Using AD2, double-click on the *OscillatorSine* icon in the design window to bring up the parameter-setting window for this CAM and enter a new frequency of 100 kHz. Download this configuration into the PAM and observe the new waveform on your oscilloscope. Select the *Sim* icon to run the simulator. Compare the simulated scope display to your actual display (adjust the V/div and Time/div to be the same between the two displays). Notice that the amplitude shown on the simulation and the actual amplitude are again different.

    Observations:

    _____

    _____

8. On the AD2 screen, click on the *Settings* menu tab. Select *Active Chip Settings* and then *Clocks* to adjust the system clock parameters. Point to the Clock 1 setting and use the LEFT/RIGHT slider to set this clock to be 40.0 kHz. (The system clock will be divided by 400.) Leave all other clock settings unchanged and click OK.

9. Double-click on the *OscillatorSine* icon to bring up the parameter-setting window. Set the parameters as follows to achieve 440 Hz (0.44 kHz) oscillation:

    (a) Point to Clock A in the upper left and use the pull down arrow to select Clock 1.

    (b) Click on the Oscillator Frequency window (temporarily RED indicating an out of range error), enter 0.44 and then click OK.

10. Add an additional wire from the output of the Inverting Gain Stage to OutputCell2. This output will be used to power the second audio channel. You have now set up the PAM as a stereo amplifier. You can observe waveforms with the oscilloscope and hear the effect with the steps that follow.

11. Plug in the power supply for the PC stereo speakers and then plug the stereo cable into the OUTPUT stereo jack next to the OUT2 and OUT1 terminal posts. Turn down the volume control to the lowest level before going to the next step.

12. Activate the *Download* icon to load the configuration file into the ASP. Turn up the volume on the stereo speakers enough to hear the 440 Hz sound. This is the piano tuning standard A note used as the reference to tune an entire piano. Adjust your oscilloscope to clearly see the waveform. Adjust the gain of the Inverting Gain Stage if needed to increase the signal amplitude.

13. Push the Analog-Reset pushbutton (labeled A-RST) near the Power Terminal Block to clear out the previous configuration file and shut down the ASP device. The green Configuration LED should turn OFF.

Conclusion: Part 2

Questions: Part 2
1. What steps are required to change the gain of the inverting amplifier?

2. What signals can the simulated oscilloscope show that the actual oscilloscope cannot show?

3. In step 7, you observed that the 100 kHz signal had a smaller amplitude from the PAM than the 50 kHz signal. What change would you make to the gain of the inverting amplifier to have a 1.0 V_p signal from the PAM?

4. What happens when the Analog-Reset pushbutton is pressed?

Experiment 13-A Basic Op-Amp Circuits

A comparator is a special operational amplifier designed as a fast switching device. It produces a high or low output, depending on which of the two inputs is larger. When the noninverting input is very slightly larger than the inverting input, the output goes to positive saturation; otherwise it goes to negative saturation. For noncritical applications, general-purpose op-amps are satisfactory and are used "open-loop" (no negative feedback). Although the application activity used a comparator, a general-purpose op-amp (the LM741C) can be used in the lab. A variation of the basic comparator, called the Schmitt trigger, will also be investigated in part 1.

In Part 2, you will investigate several applications of a summing amplifier. This amplifier can combine multiple inputs while still maintaining isolation between them. One circuit that you will investigate is a precision full-wave rectifier. The full-wave circuit contains a precision inverting half-wave rectifier and a summing amplifier.

In Part 3, you will investigate two other nonlinear circuits that have application in waveform generation and signal processing – the integrator and the differentiator. A true integrator produces an output voltage that is proportional to the *integral* (sum) of the input voltage waveform over time. The opposite of integration is differentiation. Differentiation means finding the rate of change. Because the integrator and differentiator circuits in this experiment use an inverting amplifier, the output is the *negative* result of true integration and differentiation.

Reading
Floyd, *Electronic Devices*, Eighth Edition, Chapter 13

Key Objectives
Part 1: Compare the input and output waveforms for comparator and Schmitt trigger circuits. Use an oscilloscope to plot the transfer curve for the circuits.

Part 2: Construct and test a digital-to-analog converter (DAC) using a summing amplifier. Apply a binary count sequence to it to form a step generator and observe the result. Test a full-wave rectifier that uses a summing amplifier.

Part 3: Construct and test integrator and differentiator circuits. Determine the response of these circuits to various waveforms.

Components Needed
Part 1: The Comparator and Schmitt Trigger
Resistor: one 100 kΩ
Two 1.0 μF capacitors

One 10 kΩ potentiometer
One LM741C op-amp

Part 2: The Summing Amplifier
Resistors: one 3.9 kΩ, one 5.1 kΩ, four 10 kΩ, one 20 kΩ
Capacitors: two 1.0 μF
Two signal diodes, 1N914 (or equivalent)
Two LM741C op-amps
One 7493A 4-bit ripple counter

Part 3: The Integrator and Differentiator
Resistors: two 1.0 kΩ, four 10 kΩ, two 22 kΩ, one 330 kΩ
Capacitors: one 2200 pF, one 0.01 μF, two 1.0 μF
Three LM741C op-amps
One 1.0 kΩ potentiometer
Two LEDs (one red, one green)

Part 1: The Comparator and Schmitt Trigger
The Transfer Curve
1. Figure 13-1 shows an inverting comparator circuit with a variable threshold determined by the potentiometer setting. Construct the circuit and set V_{REF} to near 0 V. Set the function generator for a 3.0 V_{pp} triangle waveform at 50 Hz and observe the input and output waveforms on a two-channel oscilloscope. Sketch the waveforms on Plot 13-1. Note the point where switching takes place. Be sure to label the axes on all plots with the voltage.

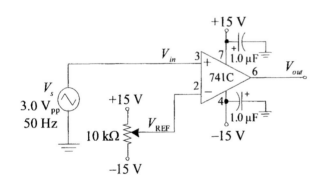

Figure 13-1

Plot 13-1 Comparator waveform.

2. Observe the output as you vary the potentiometer. Then reset V_{REF} to 0 V and change the input to a sine wave as was done in the Application Activity.

Observations:_____

3. In this step, you will plot the transfer curve for the comparator on an oscilloscope. Restore the signal to a triangle input. Place V_{in} on the X-channel and V_{out} on the Y-channel. Set the VOLTS/DIV control so that both signals are on the screen. Neither channel should be inverted. Then switch the oscilloscope to the X-Y mode. Sketch and the transfer curve you see in Plot 13-2.[1] Label the axes of the curves, showing the voltage scale for each axis.

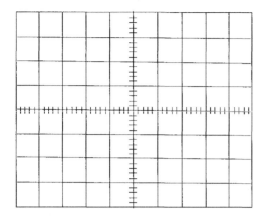   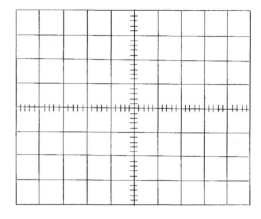

Plot 13-2 Comparator transfer curve.   **Plot 13-3** Comparator transfer curve.
(inputs reversed).

4. Vary the potentiometer as you observe the transfer curve.

Observations:_____

5. While observing the transfer curve, reverse the inputs to the comparator. Sketch the new transfer curve in Plot 13-3.

Schmitt Trigger and Its Transfer Curve
6. Construct the Schmitt trigger circuit shown in Figure 13-2. Set the potentiometer to the maximum resistance and put the oscilloscope in normal time base mode (not X-Y mode). Slowly reduce the resistance of the potentiometer and observe the input and output waveforms. Note that when the output changes states, the input voltage is different for a rising and a falling signal. Sketch the observed input and output waveforms in Plot 13-4. Label the axes of the curves.

---

[1] You may observe a slight difference in the switching point, depending on whether the input rises or falls, due to the slew rate limitation of the op-amp. Slowing the generator will reduce the effect.

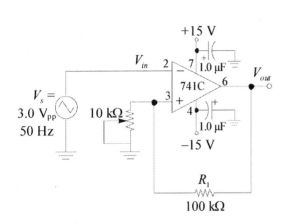

Figure 13-2

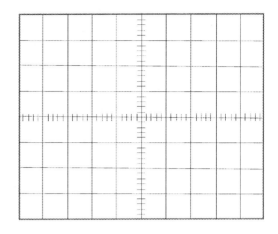

Plot 13-4 Schmitt trigger waveform.

7.   Plot the transfer curve for the Schmitt trigger on the oscilloscope. The input signal is again on the X-channel and the output signal is on the Y-channel. Select the X-Y mode and adjust the controls to view the transfer curve. Notice the hysteresis. Sketch the transfer curve you see in Plot 13-5. Label the axes of the curves.

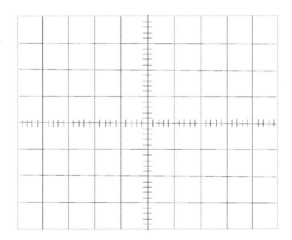

Plot 13-5 Schmitt trigger transfer curve.

8.   While observing the transfer curve, vary the potentiometer.

Observations:_____

Conclusion: Part 1

Questions: Part 1
1. Describe how the threshold voltage changes the transfer curve for a comparator.

2. Assume the circuit in Figure 13-2 had V_{REF} set to zero volts. How would you expect the output to be affected by varying the dc offset control on the generator?

3. Would a sinusoidal input to the comparators produce the same transfer curve as a triangle waveform? Explain.

4. Summarize the important differences between a comparator and a Schmitt trigger.

Part 2: The Summing Amplifier
DAC and Step Generator
1. Measure and record the values of the resistors listed in Table 13-1.

2. The circuit shown in Figure 13-3 is a summing amplifier connected to the outputs of a binary counter. The counter outputs are weighted differently by resistors R_A through R_C, and added by the summing amplifier. The resistors and summing amplifier form a basic DAC. Note that the 7493A counter is powered from a +5.0 V supply. The input to the 7493A is a logic pulse (approximately 0 to 4 V) at 1.0 kHz from a function generator. Construct the circuit. Observe V_{out} from the 741C. You should observe a series of steps. Sketch the output in Plot 13-6. Label the voltage and time on your plot.

Table 13-1

Resistor	Listed Value	Measured Value
R_A	20 kΩ	
R_B	10 kΩ	
R_C	5.1 kΩ	
R_f	3.9 kΩ	

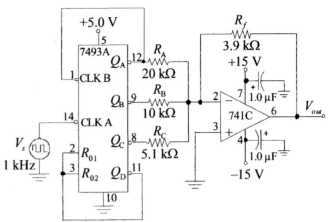

Figure 13-3

3. To see how the steps are formed, observe the Q_A, Q_B, and Q_C outputs from the 7493A. To observe the timing, put Q_C on channel 1 of your oscilloscope; trigger the scope from this channel (the *slower* signal). Keep channel 1 in place while moving the channel 2 probe. Sketch the waveforms in the correct time relation in Plot 13-7.

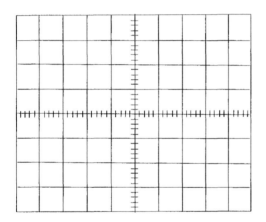

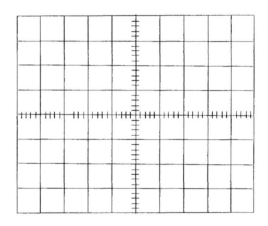

Plot 13-6   **Plot 13-7**

Precision Noninverting Half-wave Rectifier

4. Construct the precision noninverting half-wave rectifier shown in Figure 13-4. Set the input for a 1.0 kHz, 5.0 V_{pp} sinusoidal wave with no dc offset. The power supply connections on this and remaining circuits are not shown explicitly – connect the LM741C as before including the bypass capacitors. Observe the output waveform. The output follows the input almost exactly except for a small "jump" on the leading edge. (This jump is more pronounced if you raise the frequency.) The jump is caused by the time required (slew rate) for the output to go from negative saturation to +0.7 V (the voltage required to turn on the diode).

Observations:_____

_____

138

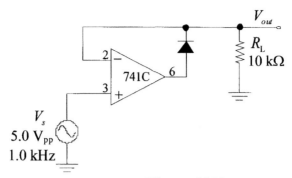

Figure 13-4

Precision Inverting Half-wave Rectifier

5. A precision inverting half-wave rectifier is shown in Figure 13-5. The diode between the op-amp output and the inverting input (D_1) prevents the output from saturating, allowing the output to change immediately after the diode starts conducting. The circuit can be recognized as an inverting amplifier with a diode added to the feedback path and the clamping diode. Construct the circuit with the input set as before and observe the output. Look at pin 6 and momentarily pull D_1 from the circuit.

Observations: _____

_____

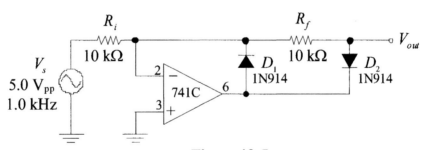

Figure 13-5

Precision Full-wave Rectifier

6. By combining the inverting half-wave rectifier from step 5 with a summing amplifier, a precision full-wave rectifier can be constructed. The circuit is shown in Figure 13-6. Construct the circuit with the input set as before. On Plot 13-8, sketch the waveforms at the left side of R_{i2} and R_{i3} (inputs) and V_{out}.

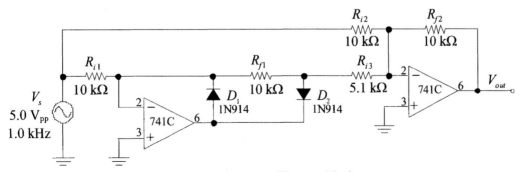

Figure 13-6

139

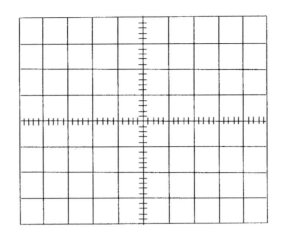

Plot 13-8

Conclusion: Part 2

Questions: Part 2

1. (a) The step generator in Figure 13-3 forms negative falling steps starting at zero volts and going to a negative voltage (approximately –4.4 V). Explain why.

    (b) How could you modify the circuit to produce positive, rising steps at the output?

2. Refer to Figure 13-3. Assume that all three inputs to the summing amplifier (Q_A, Q_B, and Q_C) are 4.5 V. Compute the output voltage from the summing amplifier.

3. Assume you have a function generator that does not have a dc offset control. Show how you could use a summing amplifier to add or subtract a dc offset from the output.

140

4. The gain for the summing amplifier in the full-wave rectifier circuit (Figure 13-6) is not the same for both inputs. Explain why.

5. The word *operational amplifier* originated from mathematical operations that could be performed with it. Assume you wanted to produce a circuit for which the output voltage was given by the expression $V_{out} = -3A - 2B$ (A and B are variable input voltages). Show how this operation could be accomplished with a summing amplifier by drawing the circuit. Show values for resistors.

Part 3: The Integrator and Differentiator

1. Construct the comparator circuit shown in Figure 13-7. Vary the potentiometer. Measure the output voltage when the red LED is on and then when the green LED is on. Record the output voltages, V_{OUT}, in Table 13-2. Then set the potentiometer to the threshold point. Measure and record V_{REF} at the threshold. It should be very close to 0 V.

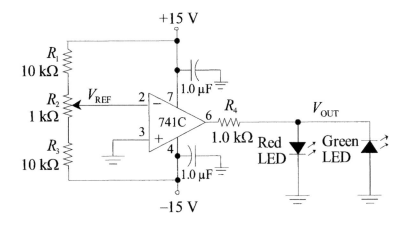

Figure 13-7

Table 13-2

V_{OUT}		V_{REF}
Red ON	Green ON	Threshold

141

2. In this step, you will test the effects of the comparator on a sinusoidal wave input and add an integrating circuit to the output of the comparator. Connect the circuit shown in Figure 13-8 with a 1.0 V_{pp} sine wave input at 1.0 kHz as illustrated. Check that there is no dc offset. Observe the waveforms from the comparator (point A) and from the integrator (point B). Adjust R_2 so that the waveform at B is centered about zero volts. Sketch the observed waveforms in the correct time relationship on Plot 13-9. Show the voltages and time on your plot.

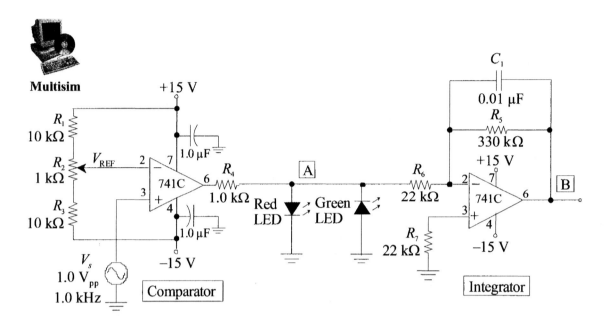

Figure 13-8

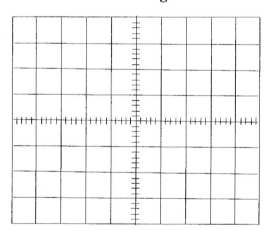

Plot 13-9

3. Vary R_2 while observing the output of the comparator and the integrator. Observations:

_____

4.   For each of the troubles listed in Table 13-3, see if you can predict the effect on the circuit. Then insert the trouble and check your prediction. At the end of this step, restore the circuit to normal operation.

Table 13-3

Trouble	Symptoms
No Negative Power Supply	
Red LED open	
C_1 open	
R_5 open	

5.   In this step, you will add a differentiating circuit to the previous circuit. The circuit is shown in Figure 13-9. Connect the input of the differentiator to the output of the integrator (point **B**). Observe the input and output waveforms of the differentiator. Sketch the observed waveforms on Plot 13-10. Show the voltages and time on your plot.

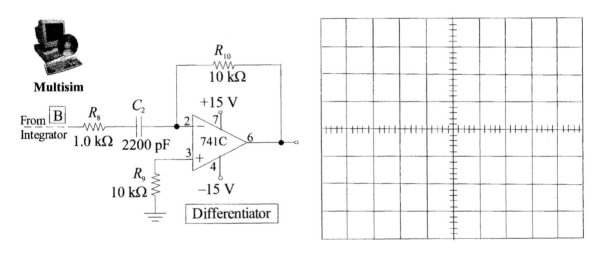

Figure 13-9

Plot 13-10

6.   Remove the input from the differentiator and connect it to the output from the comparator (point **A**). Observe the new input and output waveforms of the differentiator. Sketch the observed waveforms on Plot 13-11. Show the voltages and time on your plot.

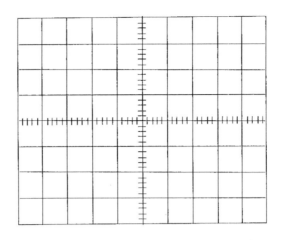

Plot 13-11

Conclusion: Part 3

Questions: Part 3

1. Compute the minimum and maximum V_{REF} for the comparator in Figure 13-7.

    $V_{REF(MIN)} =$ _____ $V_{REF(MAX)} =$ _____

2. The comparator output in Figure 13-7 did not go near the power supply voltages. Explain why not.

3. (a) For the integrator circuit in Figure 13-8, what is the purpose of R_5?

    (b) What happened when it was removed?

4.   What type of circuit will produce leading-edge and trailing-edge triggers from a square wave input?

Multisim Simulation

Multisim

Multisim files for the lab manual are on the website www.prenhall.com/floyd. Open the Multisim file Experiment_13_Integrator-nf. The circuit is the same as the one in Figure 13-8 but has switches to turn on and off power for troubleshooting. Compare the simulation results to the experimental result. Then open the file Experiment_13_Integrator-fault, which is the same file but with a fault. By observing the circuit action, you should be able to decide the likely problem, turn off power, and check it using the ohmmeter.

Another Multisim file, named Experiment_13_Int+Diff-nf, has the combination of comparator, integrator, and differentiator all together as a complete circuit. This circuit is the combination of Figures 13-8 and 13-9. It is set up with the four-channel oscilloscope so that you can view four points simultaneously. Notice how the scope is triggered. Compare your experimental results with the simulation results in this file.

Experiment 13-B Programmable Analog Design

This experiment has three independent parts that can be done in any order. In Part 1, you will investigate the comparator. A comparator is a special-purpose amplifier designed to compare two signals and produce a high or low output, depending on which of the two inputs is larger. When the noninverting input is slightly larger than the inverting input, the output goes to the positive maximum; otherwise, it goes to the negative maximum. The comparator CAM used with Anadigm ASP has three similar operating modes. It can:

- Compare two inputs
- Compare one input to ground
- Compare one input to a reference voltage

As explained in the *Electronic Devices* text, is it often desirable to have a controlled amount of hysteresis when using a comparator; this circuit is sometimes referred to as a Schmitt trigger. You can add selected amounts of hysteresis as a parameter-setting option to the comparator CAM.

In Part 2, you will investigate the summing amplifier and peak detector. A summing amplifier has two (or more) input signals that are combined. For this part, the signals are generated within the PAM, with slightly different frequencies. The output of the summing amplifier shows an interesting pattern on the oscilloscope that looks like AM modulation, but in fact is not. This signal is then processed with a peak detector that will follow the envelope. By connecting speakers to the PAM, you can listen to this signal and see the effects of small changes in the frequency of one of the sources.

In Part 3, you will test a differentiator circuit. This circuit has application in waveform generation and signal processing, including generation of triggers for digital circuits and other applications. A differentiator circuit produces an output that is proportional to the instantaneous rate of change of the input waveform. The amplitude from the differentiator is small, so amplifiers are added to the configuration. Finally, you will test a rectifier connected to the amplifier output.

Reading

Floyd, *Electronic Devices*, Eighth Edition, Chapter 13, with emphasis on the Programmable Analog Design section

Key Objectives

Part 1: Compare the input and output waveforms for a comparator and for a comparator with hysteresis. Plot the transfer curve for each circuit.

Part 2: Describe the result of summing two sine waves of nearly the same frequency and determine the effect of peak detecting the resulting signal.

Part 3: Test a differentiator circuit to determine the response of the circuit to various input waveforms.

Components Needed

Part 1: The Comparator and Comparator with Hysteresis
Programmable Analog Module
Function generator with BNC-to-grabber tip connectors

Part 2: The Summing Amplifier and Peak Detector
Programmable Analog Module
Set of self-powered stereo PC speakers, each with a 3.5 mm stereo plug

Part 3: The Differentiator
Programmable Analog Module
Function generator with BNC-to-grabber tip connectors

Part 1: The Comparator and Comparator with Hysteresis
Input Signals for the Programmable Analog Module

1. The Programmable Analog Module (PAM) has a number of selector switches that are provided to enable different operating modes. The PAM is designed to accept either differential input signals or the single-ended signals more commonly seen in everyday experience. The 10-position Input Termination DIP switch, labeled SW6, located along the edge of the circuit board between the INPUT and OUTPUT stereo jacks is used to make this selection. Set 1, 5, 6, and 10 to ON and all others to OFF to use the PAM for single-ended input signals.

2. Set up the function generator prior to connecting it to the circuit by viewing the output of the generator on an oscilloscope. Set the function generator for a 6.0 V peak-to-peak triangle waveform at 50 Hz with zero offset. The PAM and the Anadigm ASP are set up for differential inputs. Because the function generator is a single-ended signal source, the internal PAM signal is 3.0 V_{pp}. Thus, the external signal observed from the function generator is *double* the internal signal. You will explore this idea further in Experiment 14B, Part 1.

3. Use the grabber clips from the function generator to connect to Input1P and the GND terminal post as shown in Figure 13-10. Two open holes at each position, one above the other, are the same signal. Notice the OUT1 and OUT2 terminal posts where the output signals will appear for these experiments.

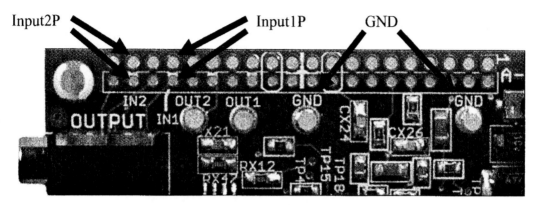

Figure 13-10 Signal input pins to PAM.

4. Start up AD2. Press the M key to open the *Insert New CAM* window and place the comparator CAM in the design window and accept the default settings shown in Figure 13-11. The default settings are:
    - Compare the input signal to ground
    - Non-inverted outputs
    - 0 mV of hysteresis

    You can ignore the Sampling and Synch settings for this experiment.

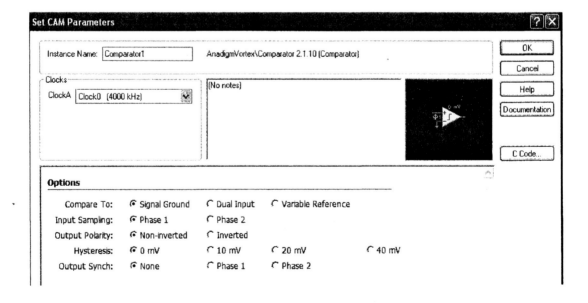

Figure 13-11 Comparator CAM parameter-settings window.

The Comparator Output and the Transfer Curve

5.   Insert a second comparator with the default settings in the design window as shown in Figure 13-12. Wire the circuit as shown and download it to the PAM. The function generator should still have the 6.0 V peak-to-peak triangle waveform that you set up in step 2. Observe the signal from the function generator and the OUT1 waveforms on your oscilloscope. Sketch the two waveforms on Plot 13-12. Be sure to label the axes and note the switching points (rising and falling).

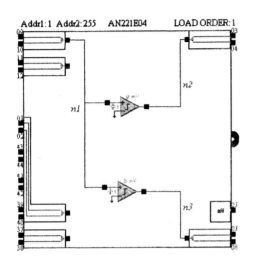

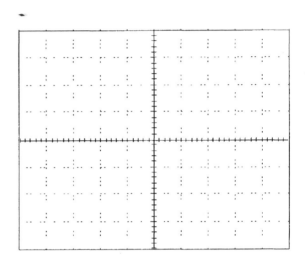

Figure 13-12 Two comparators.   **Plot 13-12**

6.   Double-click on the upper comparator icon to open the parameter-setting window and select the *Variable Reference* option. Accept the +1 V default value, close the window, and download into the PAM. Observe and sketch the new OUT1 waveform on Plot 13-13. Label the plot and the axes.

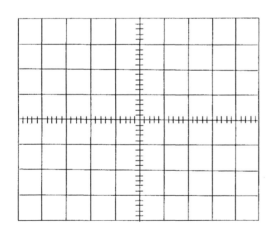

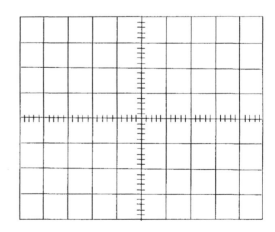

Plot 13-13   **Plot 13-14**

7. Change the variable reference value to −1 V and download into the PAM. Observe and sketch the new OUT1 waveform on Plot 13-14.

8. Change the variable reference value to 0 V and download this into the PAM. Adjust the dc offset from the function generator to +1 V and then to −1 V. What did you observe about the OUT1 waveform from the PAM in each of these cases?

   Observations:_____

   _____

9. In this step, you will plot the transfer curve for the comparator on an oscilloscope. This is a plot of the output voltage versus the input voltage. Restore the function generator to 0 V offset with the 6.0 V peak-to-peak triangle waveform at 50 Hz (*Reminder* – this is 3.0 V_{pp} internal to the PAM). Connect the X-channel of the scope (typically Channel 1 but this is model specific) to Input1P of the PAM and the Y-channel to OUT1 of the PAM. Adjust the oscilloscope so that both signals are on the screen. Neither channel should be inverted. Then switch the oscilloscope to the X-Y mode. Sketch the transfer curve you see in Plot 13-15. Label the axes of the curves, showing the voltage scale for each axis.

Plot 13-15 Comparator transfer curve.

10. Change the variable reference voltage to +1 V and download the configuration file to the PAM.

    Observations:_____

11. Restore the variable reference voltage to 0 V and download the configuration file. Vary the offset voltage of the function generator.

    Observations:_____

The Comparator with Hysteresis and Its Transfer Curve

12. To see the effect of hysteresis on a comparator, reduce the signal from the function generator to 1.0 V$_{pp}$. Select the lower comparator CAM in Figure 13-12 and select the largest hysteresis available, which is 40 mV. Select Signal Ground as the reference (0 V) for both comparators. Put the oscilloscope in normal time base mode (not X-Y mode). Download the new configuration file into the PAM. Show the input and OUT2 on the oscilloscope. Notice that the output switches states for a different input voltage depending on if it is a rising signal or a falling signal. Sketch the input and output waveforms in Plot 13-16. Label the axes.

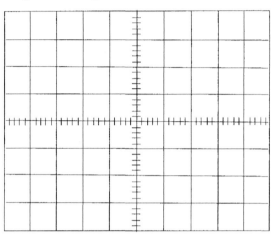

Plot 13-16 Comparator with hysteresis waveform.

13. Plot the transfer curve you see on the oscilloscope. The input signal is again on the X-channel and the output signal is on the Y-channel. Select the X-Y mode and adjust the controls to view the transfer curve. If you have trouble seeing hysteresis on a digital scope, try turning on Persist on the display. Sketch the transfer curve you see in Plot 13-17. Label the plot and axes.

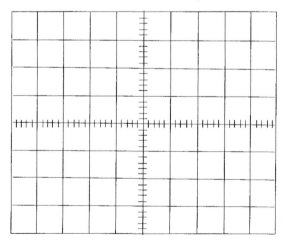

Plot 13-17 Schmitt trigger transfer curve.

Conclusion: Part 1

Questions: Part 1

1. Describe the difference between a differential signal and a single-ended signal.

2. Describe in your own words the meaning of hysteresis.

3. What is the purpose of hysteresis for a comparator?

4. This experiment uses a triangle wave as the input. If the input signal is changed to a sine wave, would the transfer curve stay the same or change? Explain your answer.

Part 2: The Summing Amplifier and Peak Detector

1. Place an inverting summing amplifier (*SumInv*) CAM in the center of the design window. In the *Set CAM Parameters* window, under Clocks, choose Clock3 at 250 kHz in the ClockA selection box and set both input gains to 0.5. You will use Clock3 in all CAMs in this Part; if you see a dotted wire, check that Clock3 is specified.

2. Place a sine wave oscillator (*OscillatorSine*) CAM to the left side of the design window. In the *Set CAM Parameters* window, choose Clock3 at 250 kHz in the ClockA selection box. Under Parameters, change the oscillator frequency to 20 kHz and the peak amplitude to 2.0 V.

3. Place a second sine wave oscillator CAM under the first oscillator. In the *Set CAM Parameters* window, choose Clock3 at 250 kHz in the ClockA selection box. Under Parameters, change the oscillator frequency to 20.5 kHz and the peak amplitude to 2.0 V.

153

4. Place a peak detector CAM in the upper right side of the design window. In the *Set CAM Parameters* window, use Clock3 for both ClockA and ClockB. Choose the decay time constant of 511 μs (maximum) and accept other defaults. Then wire the circuit as shown in Figure 13-13.

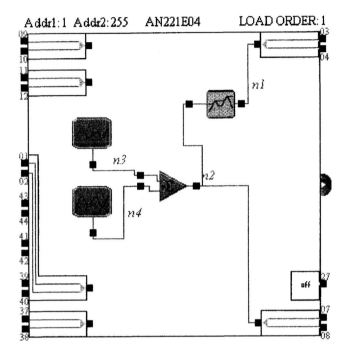

Figure 13-13 Two oscillators, summing amplifier, and peak detector.

5. In the AD2 design window, place a scope probe on the output of the summing amplifier and the output of the peak detector. In the *Simulate* menu, choose Setup Simulation and choose a Start Time of 0 (default) and an End Time of 10 ms (10 m). Run the simulation. Set the scope Time Per Division to 1 ms.

Observations:_____

6. Plug in the power supply for the PC stereo speakers and then plug the stereo cable into the OUTPUT stereo jack (next to the OUT2 terminal post). Turn down the volume control to the lowest level before going to the next step.

7. Download the configuration file to the PAM. Connect a scope to the posts for OUT1 and OUT2. It is tricky obtaining a stable display because of the multiple trigger points. Sketch the outputs in Plot 13-18 and compare them to the simulation. Label the axes of your plot.

Observations:_____

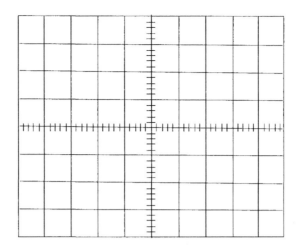

Plot 13-18 Summing amplifier and peak detector outputs.

8. You may want to try experimenting with different frequencies of the oscillators. Try a few different combinations and note the effect on the waveforms and the sound in the speakers.

    Observations:_____

9. Delete the peak detector in the design window. Add a wire directly from the summing amplifier to OutputCell1. Download this configuration to the PAM with the speakers still connected.

    Observations:_____

Conclusion: Part 2

Questions: Part 2

1. What is the purpose of the decay time constant for the peak detector? How would a different time constant affect the output?

2.   In step 9, you tested the circuit without the peak detector. Why do you think there was a significant difference in the result?

3.   What general statement can you make about the frequency from the peak detector?

Part 3: The Differentiator

1.   Place an inverting differentiator (*Differentiator*) CAM in the upper middle area of the design window. Set the Differentiation Constant to 0.1 μs (shown in AD2 as 0.1 u) and accept the other defaults. Connect the input to InputCell1 and the output to OutputCell1.

2.   Add two inverting gain stages (*GainInv*) below the differentiator and set the gain for each to be 12. The differentiator produces a tiny signal, so the two stages together will produce a noninverting gain of 144. Wire the circuit as shown in Figure 13-14, with the output of the gain stages going to OutputCell1.

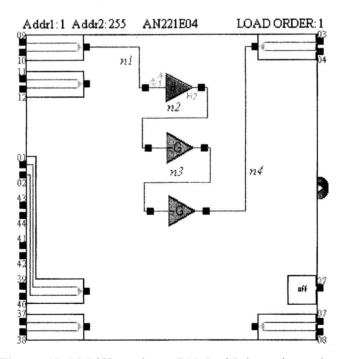

Figure 13-14 Differentiator CAM with inverting gain stages.

3. Set up the function generator for a 5 V_{pp} triangle waveform at 20 kHz with no dc offset. Observe the input waveform from the function generator and the output of the circuit on OUT1 (following the gain stages). Sketch the observed waveforms on Plot 13-19. Label the axes of your plot.

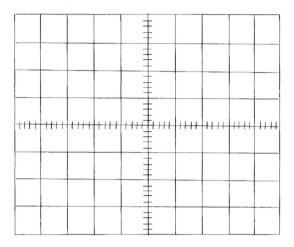

Plot 13-19 Differentiated output triangle wave.

4. Adjust the dc offset on the function generator slightly positive and slightly negative.

   Observations:_____

5. Change the function generator to a sine wave output with no offset. Observe the new input and output waveforms of the differentiator. Sketch the observed waveforms on Plot 13-20. Notice the phase shift between input and output.

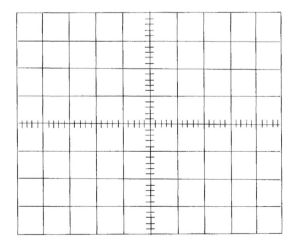

Plot 13-20 Differentiated output of a sine wave.

157

6. Change the function generator to a square wave output with no offset. Observe the new input and output waveforms of the differentiator. Sketch the observed waveforms on Plot 13-21.

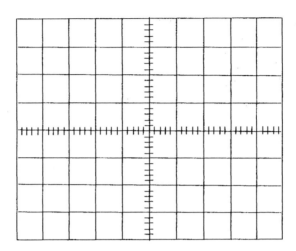

Plot 13-21 Differentiated output of a square wave.

7. Add a rectifier filter (*RectifierFilter*) CAM in the lower part of the design window. In the *Set Cam Parameters* window, choose a positive half-wave rectifier. Then set the Corner Frequency to 400 kHz and the Gain to 2. Connect the output of the second gain stage to the input and the output of the rectifier to OutputCell2 as shown in Figure 13-15. Compare the signals at OutputCell1 and OutputCell2. You may need to change the trigger source to channel 2 to get a stable display. Sketch the waveform at OutputCell2 in relation to the square wave input.

Observations:_____

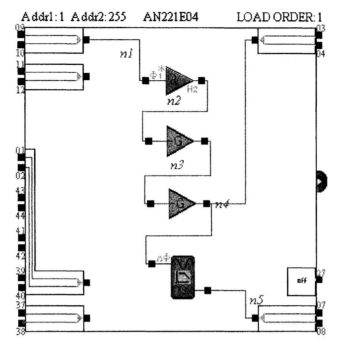

Figure 13-15 Differentiator CAM with inverting gain stages and rectifier

Conclusion: Part 3

Questions: Part 3

1. In step 4, you varied the dc offset. Why do you think this had very little effect on the output?

2. If you differentiate a triangle waveform, does the input frequency affect the amplitude of the output? Why or why not?

3. When a sine wave is differentiated, what is the shape of the output?

Experiment 14-A Special-Purpose Op-Amp Circuits

An important circuit for amplifying low-level signals in the presence of noise is the instrumentation amplifier (IA). In Part 1, you will test an IA that you construct from three op-amps, similar to the one in Figure 14-2 of the text. The same basic circuit is used in the Application Activity except the AA uses an integrated IA. A 555 timer (discussed in text Chapter 16) is battery-powered for isolation and will simulate a low-level signal differential-input source for testing. You will be able to see how effective the IA is for rejecting simulated noise while passing the desired signal.

Another processing problem occurs with large dynamic range signals for A/D conversion, transmission, and recording and similar cases. For such signals, the dynamic range can be compressed with a logarithmic amplifier (*log amp*). A log amp is a nonlinear amplifier that converts a large change in the input signal to a small change in the output signal. In Part 2, you will test a log amp and antilog amp constructed from the LM741C.

Reading

Floyd, *Electronic Devices*, Eighth Edition, Chapter 14

Key Objectives

Part 1: Show how an instrumentation amplifier (IA) can selectively amplify a small differential signal in the presence of a large common-mode signal.

Part 2: Construct log and antilog amplifiers and observe signal compression and reconstruction.

Components Needed

Part 1: The Instrumentation Amplifier

Resistors: one 22 Ω, one 470 Ω, two 1.0 kΩ, one 8.2 kΩ, five 10 kΩ, three 100 kΩ

Capacitors: one 0.01 μF, two 1.0 μF

Three 741C op-amps

One 555 timer IC

One small 9 V battery

One 5 kΩ potentiometer

30 cm twisted-pair wire

Part 2: The Log Amplifier and Antilog Amplifier

Resistors: two 100 kΩ

Capacitors: one 0.01 μF, two 1.0 μF

Two 1N914 signal diodes (or equivalent)

Two 741C op-amps

Two 2N3904 *npn* transistors (or equivalent) (*Note:* You will obtain best results if ßs match).

Part 1: The Instrumentation Amplifier

1. Measure and record the values of the resistors listed in Table 14-1. For best results, R_1 and R_2 should match.

2. Construct the circuit shown in Figure 14-1. This is an instrumentation amplifier (IA) similar to the Application Activity except it is constructed from separate op-amps. There is a single gain-setting resistor named R_G. Also, one of the 10 kΩ resistors has been replaced with a fixed resistor and potentiometer to compensate for small circuit variations that occur with fixed resistors.

Table 14-1

Resistor	Listed Value	Measured Value
R_1	10 kΩ	
R_2	10 kΩ	
R_G	470 Ω	
R_3	10 kΩ	
R_4	10 kΩ	
R_5	10 kΩ	
R_6	8.2 kΩ	
R_8	100 kΩ	
R_9	100 kΩ	

The IA is shown in the dotted box. It is driven by the generator in differential mode. The purpose of R_8 and R_9 is to assure a bias path for the op-amps when the source is isolated later in the experiment (step 7). To simplify the schematic, the power supply connections and bypass capacitors are not shown. (Use ±15 V for all op-amps; use two 1.0 μF bypass capacitors installed near one of the op-amps.) As shown, R_9 is shorted, but it will not be shorted in step 7.

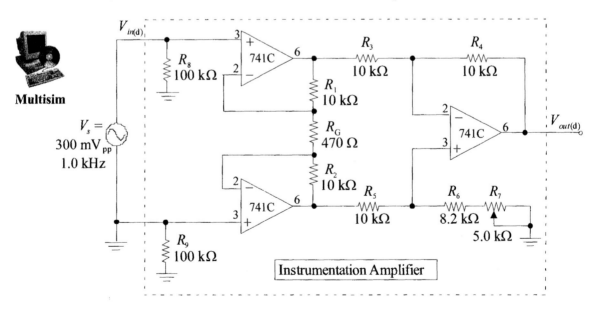

Figure 14-1 Circuit to measure differential parameters.

3. Set the potentiometer (R_7) in the middle of its range. ($R_6 + R_7$ should add to approximately 10 kΩ.) Set the input for a 300 mV$_{pp}$ sine wave at 1.0 kHz and measure this signal. This represents the differential-mode input signal, $V_{in(d)}$. Compute the differential gain from the equation:

$$A_{v(d)} = 1 + \frac{2R_1}{R_G}$$

Use this result to calculate the differential output voltage $V_{out(d)}$. Record both the computed $A_{v(d)}$ and $V_{out(d)}$ in Table 14-2. Then measure and record the input and output voltages and measured differential gain in the last column of the table.

Table 14-2

Step	Parameter	Computed Value	Measured Value
	Differential input voltage, $V_{in(d)}$	300 mV$_{pp}$	
3	Differential gain, $A_{v(d)}$		
	Differential output voltage, $V_{out(d)}$		
	Common-mode input voltage, $V_{in(cm)}$	10 V$_{pp}$	
4	Common-mode gain, $A_{v(cm)}$		
	Common-mode output voltage, $V_{out(cm)}$		
5	CMRR′		

4. Drive the IA with a common-mode signal as shown in Figure 14-2. Set the signal generator for a 10 V$_{pp}$ signal at 1.0 kHz ($V_{in(cm)}$) and measure this signal. Observe the output voltage and adjust R_7 for *minimum* output. Measure the peak-to-peak output voltage, $V_{out(cm)}$. Determine the measured common-mode gain, $A_{v(cm)}$, by dividing the measured $V_{out(cm)}$ by the measured $V_{in(cm)}$. Record the data in Table 14-2.

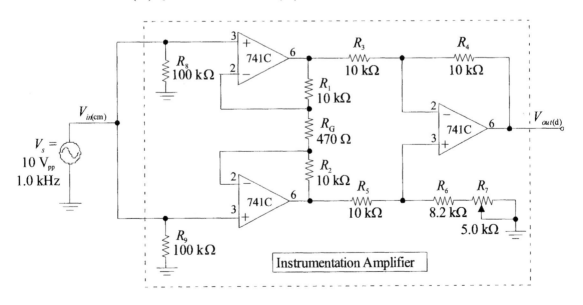

Figure 14-2 Circuit to measure common-mode parameters.

163

5. Calculate the CMRR' (in dB) from the equation CMRR' = 20 log $A_{v(d)}/A_{v(cm)}$. Because it is based on measured gains, enter this as the *measured* value in Table 14-2. When you have completed the common-mode measurements, reduce the signal generator frequency to 60 Hz. This will represent the (undesired) noise source which will be the common-mode signal for the next section.

Adding Differential-mode and Common-mode Sources

6. In this step, you will build a pulse oscillator to serve as a small differential signal source in the presence of electrical noise. The oscillator is constructed from a 555 timer. Construct the circuit shown in the shaded box in Figure 14-3 (preferably on a separate protoboard if you have one available). R_C and R_D serve as an output voltage divider to reduce the signal to a small value (such as you might find from a transducer). Measure the output frequency and voltage and indicate these values in the first two rows of Table 14-3.

   Note that the differential signal source must be *floating* (no common ground with IA) so it is powered by a small 9 V battery as shown.

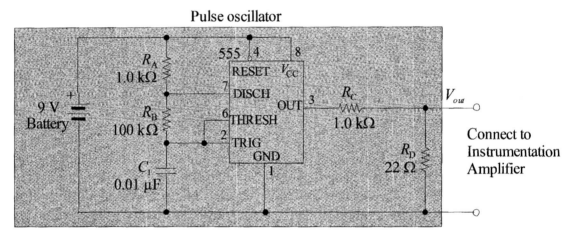

Figure 14-3 Oscillator to serve as a source for the instrumentation amplifier.

Table 14-3

Parameter	Measured Value
Oscillator frequency	
$V_{out(pp)}$ from oscillator	
$V_{out(pp)}$ from IA	

7. Connect the oscillator to the IA as shown in Figure 14-4 with about 30 cm of twisted-pair wire to simulate a short transmission line. Be sure there is no common ground from the oscillator to the input of the instrumentation amplifier. Measure the output signal from the IA and record the output in the last row of Table 14-3. Note that this signal represents a differential-mode input and is therefore amplified by the differential gain, $A_{v(d)}$.

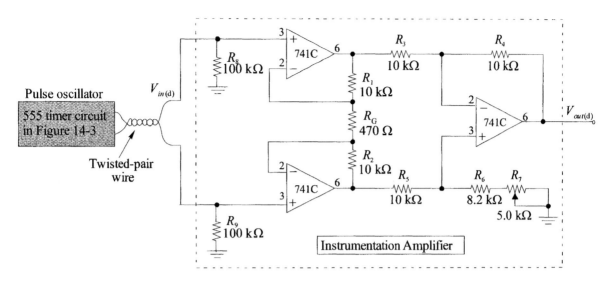

Figure 14-4 Connecting the oscillator to the instrumentation amplifier.

8.   Add a simulated source of common-mode noise to the oscillator. Often, the noise source is 60 Hz power line interference, but seldom is it as large or as well connected. Set up your function generator for a 10 V_{pp} sine wave at 60 Hz to simulate a large amount of common-mode power-line interference. Connect the generator to one side of the oscillator as shown in Figure 14-5. Observe the output signal from the IA. Adjust R_7 for *minimum* common-mode signal.

Observations: _____

_____

_____

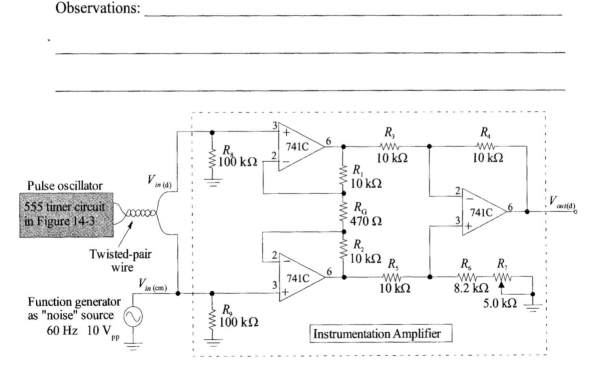

Figure 14-5 Adding a common-mode "noise" source to the oscillator.

Conclusion: Part 1

Questions: Part 1

1. Some instrumentation amplifiers have a CMRR' of 130 dB. Assuming the circuit in this experiment had a CMRR' of 130 dB with the same differential gain, what common-mode output would you expect?

2. How was the IA able to pass the oscillator signal while simultaneously blocking the signal from the function generator?

3. Why was it important to power the 555 timer using a separate battery rather than using the same supply that powered the op-amps?

4. What advantage does an IA, such as the one that you constructed, have over an ordinary differential amplifier?

Part 2: The Log Amplifier and Antilog Amplifier

1. Measure and record the values of the resistors listed in Table 14-4.

Table 14-4

Resistor	Listed Value	Measured Value
R_1	100 kΩ	
R_2	100 kΩ	

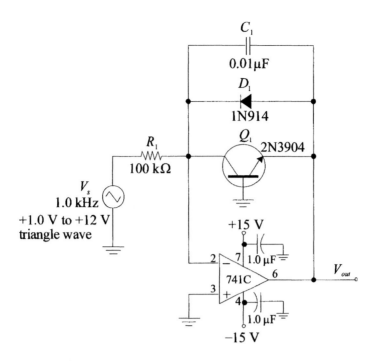

Figure 14-6

2. Construct the log amp shown in Figure 14-6. The log amp shown is designed *only* for positive signals at the input. Before connecting the function generator to the circuit, set the signal to a 1.0 kHz, +1.0 V to +12 V positive triangle waveform. Adjust the generator's dc offset control to achieve a positive waveform.

3. Observe the waveform at the output of the log amp. Sketch the input and output waveforms on Plot 14-1. Label the voltage and time on the plot. Notice the polarity of the output.

4. Change the input waveform to a positive sine wave (+1.0 V to +12 V). Sketch the input and output waveforms on Plot 14-2. Label the voltage and time on the plot.

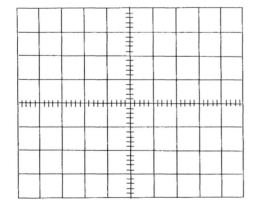

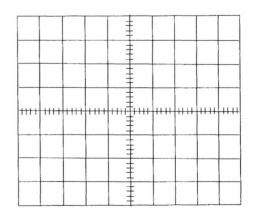

Plot 14-1   **Plot 14-2**

5. Connect the antilog amp shown in Figure 14-7 to the output of the log amp. The power inputs (not shown) should be connected to ±15 V. If the circuit is working correctly, you should see that the output waveform from the antilog amp matches the input to the log amp. Check both the sine wave and the triangle wave you tested earlier.

    Observations:_____

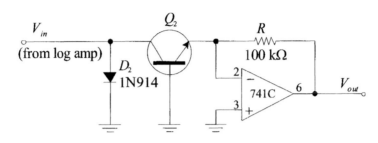

Figure 14-7

6. Check the temperature sensitivity of the circuit by grasping the transistor on the log amp between your fingers and note what happens to the output from the antilog amp. Try the same test on the diode on the antilog amp.

    Observations:_____

7. To obtain quantified data on the log amp, replace the function generator with a positive dc power supply. Set the supply to each voltage listed in Table 14-5 and measure and record the output using a DMM. Plot the results on the semilog plot in Plot 14-3.

Table 14-5 Data for Log Amp

V_{IN}	V_{OUT}
+1.0 V	
+2.0 V	
+4.0 V	
+6.0 V	
+8.0 V	
+10.0 V	
+12.0 V	

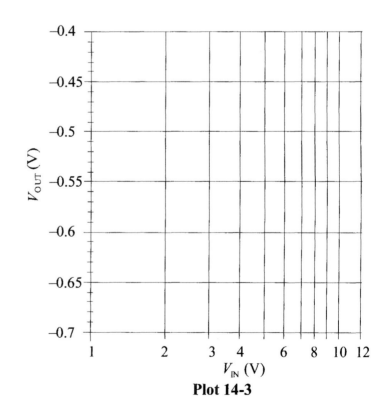

Plot 14-3

168

Conclusion: Part 2

Questions: Part 2

1. Cite at least three problems with the log amp in this experiment that are improved in commercial IC log amps.

2. Plot 14-3 represents the transfer curve for the log amp. (Recall that a transfer curve is the output plotted against the input.) Describe the transfer curve for the antilog amp. Should it be plotted on the same type of paper to obtain a straight line response?

3. The antilog amp in the experiment used the same value resistor as the log amp. What would happen if they were different? (Can you obtain gain this way?)

4. Figure 14-8 illustrates a basic arithmetic application for log - antilog amplifiers. Assume that all amplifiers invert their inputs and the summing amplifier has unity gain (-1).
    (a) What arithmetic operation is performed?

    (b) Assuming proper calibration, what is V_{OUT}?_____

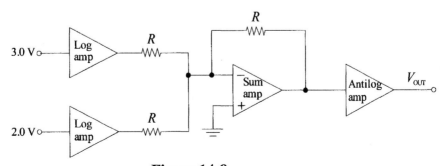

Figure 14-8

169

Multisim Simulation

Multisim

Multisim files for the lab manual are on the website www.prenhall.com/floyd. The basic instrumentation amplifier from Part 1 is simulated in Multisim but with different input signals. The circuit file is named Experiment_14_IA-nf. One of the instruments connected to the circuit is the spectrum analyzer, a very useful instrument that allows you to view signals in the frequency domain and shows the very small common-mode signal in the presence of the large differential signal. It can do this because of logarithmic view capability. The Multisim spectrum analyzer requires a minimum of a 1 kHz input, so the two signals are different frequencies than in the experiment. Even though the common-mode signal is 50 times larger at the input, it is not seen on the oscilloscope because it is attenuated by the CMRR. The spectrum analyzer shows the common mode signal is −40 dB below the differential signal when the potentiometer (R_8 in Multisim) is adjusted correctly. Notice that other signals are shown on the spectrum analyzer. Can you figure out how they are generated? This is an excellent simulation to learn about the spectrum analyzer, so take the time to adjust it and see the effects of your changes.

Experiment 14-B Programmable Analog Design

Instrumentation amplifiers (IAs) are special amplifiers designed to amplify differential signals and reject common-mode signals. They are widely used in measurement circuits. The Analog Signal Processor (ASP) also processes signals internally using differential circuits because of low drift and low noise characteristics. You probably are more familiar with single-ended signals because they are more common in instruments like the function generator, but it is important to understand both types of signals. The term "single-ended" means a signal driven by an amplifier that has a single output pin; these signals are typically referenced to ground. In Part 1, you will set up and measure a single-ended signal using the oscilloscope and then apply it to the input of a PAM. The PAM and the Anadigm ASP process signals internally in differential mode (a result that you will see in Part 2). After measuring the exact gain, you will adjust the gain within the ASP to achieve exact one-to-one relation between a single-ended input and an output from the PAM. You can use the result in this Part if you need to set an exact gain in the future.

In Part 2, you will be introduced to two new CAMs that are useful special-purpose circuits that have wide applicability in instrumentation and measurement circuits. In addition, the peak detector CAM, used in Experiment 13B, will be tested in conjunction with a full-wave rectifier to see how the decay time constant affects the signal. The new CAMs are:

- Zero-Crossing Detector CAM - outputs a digital signal when the input analog signal becomes greater than a reference voltage.
- Limiter CAM – amplifies the input signal up to a specified output limit value and then holds constant until the input reduces to where the output limit no longer applies.

Part 2 ends with another look at the comparator CAM with hysteresis (first given in Experiment 13B). The comparator is useful in instrumentation, so the application using the tank simulation in the text is described using a function generator as the input instead of a pressure sensor. In the laboratory, it is difficult to obtain a precise source with the tiny signal from a simulated pressure sensor, so the function generator is set for a larger input signal and the gain stages used in the text are eliminated. You should be able to test this in a few minutes and see that the simulation and implementation are in good agreement.

Reading

Floyd, *Electronic Devices*, Eighth Edition, Chapter 14, including review of the section on Programmable Analog Design.

Key Objectives

Part 1: Test the effect of a single-ended signal when it is applied to a differential circuit. Set the gain of an amplifier CAM to calibrate the PAM.

Part 2: Test several special-purpose CAMs:
- Zero-Crossing Detector CAM
- Peak Detector and Rectifier
- Limiter CAM
- Comparator with hysteresis

Components Needed

Part 1: Single-Ended Signals into a Differential Signal Circuit
Programmable Analog Module
Function generator with BNC to grabber tip connectors

Part 2: Instrumentation CAMs
Programmable Analog Module
Function generator with BNC to grabber tip connectors

Part 1: Single-Ended Signals into a Differential Signal Circuit

1.   For this part, you will need to have the PAM connected to the PC and AnadigmDesigner2 running. Carefully set your function generator for a 5 kHz, 1.0 V_{pp} triangle wave output with zero offset. Using your grabber clips connect the ground of the function generator to the ground on the PAM and the output of the function generator to Input1P on the PAM (see Figure 13-10). Connect channel-1 of your scope to observe the signal from the function generator.

2.   Open AnadigmDesigner2. In the design window, place a gain limiter (*GainLimiter*) CAM in the upper center area. In the *Set CAM Parameters* window, leave the gain at 1.0 but change the *Output Voltage Limit* to 4.0 V. Connect a straight-through wire from InputCell1 to OutputCell1 and a second wire from InputCell1 to the input of the gain limiter stage. Connect the output of the gain limiter stage to OutputCell2. Your design window should look like Figure 14-9.

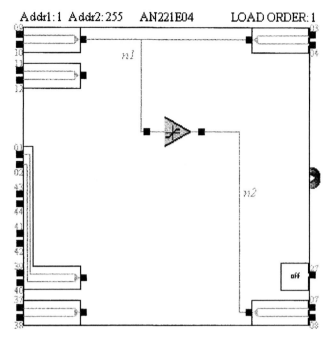

Figure 14-9 Circuit to explore single-ended and differential signals.

3.   Check the voltage again from the function generator to ensure there was no drift
     or loading effect, as you need a value that is as accurate as possible. Record the
     measured peak-to-peak voltage from the function generator while it is connected
     to the PAM and this voltage divided by 2.

     (a) Function generator peak-to-peak voltage: _____

     (b) Function generator peak-to-peak voltage divided by 2: _____

4.   Move your scope probe to the OUT1 terminal on the PAM and measure the peak-
     to-peak voltage of the output signal.

     OUT1 peak-to-peak voltage: _____

The OUT1 signal is approximately the same
as you saw in step 3(b). The reason is that
the Anadigm ASP is a differential-input and
differential-output device that is shown
schematically with polarity indicators.
Figure 14-10 shows the idea. Because one of
the differential inputs is grounded when a
single-ended input is used, the delivered
signal is only half as large as a differential
signal would be.

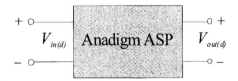

Figure 14-10 Differential input
and output signals.

5. The measured output signal at OUT1 is probably a little different than the voltage recorded in step 3(b). It is common for analog circuits not to achieve precise gain values. Typically this is of little consequence because the user can readily change the gain if necessary. Using the same scope probe as in step 4, measure and record the amplified signal at OUT2.

    OUT2 peak-to-peak voltage: _____

6. In the *Set CAM Parameters* window, adjust the gain of the gain limiter CAM so that OUT2 is the same as OUT1. The required setting is the ratio of the result in step 5 to the result in step 3(b) and will be close to 1. (You might need to adjust this a little).

    Gain setting to equalize outputs: _____

7. Change the gain to twice that of step 6 to achieve the full amount of output signal at OUT2 as recorded in step 3(a). Make any additional gain adjustment to achieve exactly 1:1 input to output voltage between the single-ended input and the differential output.

    Gain setting to set single-ended input equal to differential output: _____

    You can use this setting if you want to set an exact value of gain with the PAM. The PAM supplies an input buffer and an output buffer to the Anadigm ASP as shown in Figure 14-11 (representative signals are shown with a direct feed-through in the ASP). The input buffer is differential-in / differential-out. The output buffer is differential-in / single-ended out.

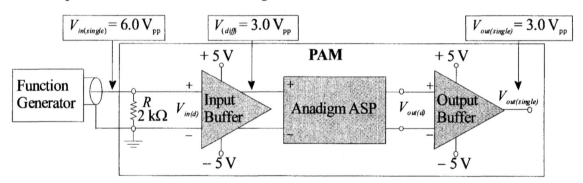

Figure 14-11 PAM input and output buffers with representative signal levels.

Conclusion: Part 1

Questions: Part 1

1. Why was it important to use the same scope probe to measure the signal amplitude of the function generator and the outputs of the PAM?

2. If a single-ended signal is transmitted over an unshielded pair of wires (signal on one wire and ground on the other wire) across a building, would you expect some amount of 60 Hz (or 50 Hz in Europe) energy to be included in the signal delivered at the far end of the pair of wires? Explain.

3. Aircraft use powerful 400 Hz electrical generators. Airframes are typically constructed of aluminum, which is a very good electrical conductor. If you wanted to send a single-ended signal from the cockpit to the tail, would it be okay to just send the signal over one wire and make the ground connections to the aluminum structure at both ends? Why or why not?

4. What might be the causes for the small differences in the actual gain achieved in a typical analog circuit?

Part 2: Instrumentation CAMs
Zero-crossing Detector

1. Place the zero-crossing detector CAM (*ZeroCross*) in the design window. In the *Set CAM Parameters* window, change ClockA to Clock3 (250 kHz). Accept the other default settings for zero-crossing detection at ground voltage with no hysteresis.

2. ` Wire the input of the zero-crossing detector to InputCell1 and the output to OutputCell2. Connect a wire from InputCell1 signal to OutputCell1 for comparison as shown in Figure 14-12. Double-click on the OutputCell1 rectangle to make sure that it is set for *Bypass* mode (this means a straight through connection). Connect the function generator to Input1P at the edge of the PAM and set it for a 2.0 V_{pp}, 1.0 kHz sine wave, with no offset. (Figure 13-10 shows the Input1P.) Download the configuration file to the PAM.

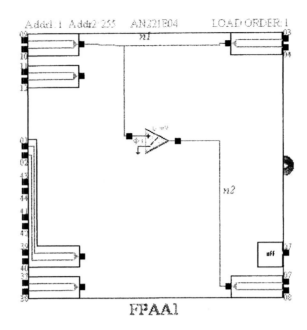

Figure 14-12 Circuit for the Zero-Crossing detector.

3.   Observe the sine wave signal at OUT1 using the oscilloscope and the output of
      the zero-crossing detector at OUT2. Sketch the waveforms below. Label the axes.

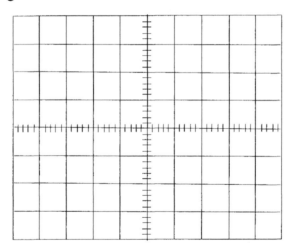

Plot 14-4 Sine wave input and zero-crossing detector output.

4.   Decision-making circuits like comparators, Schmitt triggers, and zero-crossing
      detectors have a small amount of delay between the moment when the input
      signals achieve the comparison criteria and the output signal is asserted. The
      delay is more significant at higher frequencies because of the shorter period of the
      signal. At 1 kHz, the delay is difficult to see. Try raising the frequency to 50 kHz,
      and look for the delay.

      Observations:_____

      _____

5. Another consideration in using a circuit like the zero-crossing detector is the clock speed. Find out what happens if the clock speed is increased. In the *Settings* menu, choose Active Chip Settings... This will open a new window with slider bars that allow you to choose clock speeds for each of four clocks. Try changing Clock3 to 1000 kHz and downloading the file. Then test 1600 kHz for Clock3 and observe what happens. After checking the result, restore Clock3 to 250 kHz.

    Observations:_____

    _____

Peak Detector

6. Open a new instance of AD2. Place a peak-detector CAM (*PeakDetect2*) in the upper center of the design window. In the *Set CAM Parameters* window, use Clock0 (4000 kHz) for both ClockA and ClockB. Set the decay time constant for 1 μs accepting all other defaults. As you will see, this small decay will have virtually no effect on the peak detection.

7. Place a duplicate peak detector in the center-right area of the design window (you can use copy and paste commands to do this). Place a rectifier-filter CAM (*RectifierFilter*) to the left of the lower peak detector, accepting defaults (full-wave, noninverting). Wire the circuit as shown in Figure 14-13. Download the configuration file to the PAM.

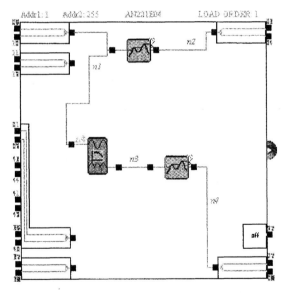

Figure 14-13 Peak detector CAMs with rectifier.

8. Connect the function generator to Input1P at the edge of the PAM and set it for a 2.0 V$_{pp}$, 1.0 kHz sine wave, with no offset. (Figure 13-10 shows the Input1P.) Observe the output at OUT1 and OUT2.

    Observations:_____

    _____

9. Double-click each of the peak detector CAMs and change ClockB to Clock3 (250 kHz) and the decay time constant to 100 μs. Download the new configuration and observe the result.

   Observations:_____

   _____

10. Increase the decay time constant on both peak detectors to the maximum (511 μs), and download the new configuration.

    Observations:_____

    _____

11. While observing one of the outputs, try raising the frequency of the function generator. (You may need to adjust your scope trigger to keep a stable display.)

    Observations:_____

    _____

Gain-Limiter

12. Open a new instance of AD2. Place an oscillator CAM (*OscillatorSine*) on the left side of the design window. In the *Set CAM Parameters* window, use Clock3 (250 kHz) for ClockA and set the frequency for 5 kHz with a peak amplitude of 2.0 V. Place a gain-limiter CAM (*GainLimiter*) in the design window. In the *Set CAM Parameters Window*, use Clock3 (250 kHz) for ClockA, accepting other defaults. Wire the circuit as shown in Figure 14-14.

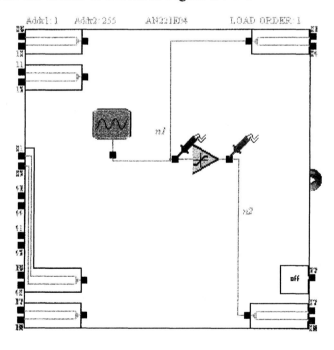

Figure 14-14 Gain limiter circuit.

13. Place the simulated probes on either side of the limiter stage as shown in Figure 14-14. In the *Simulate* menu, choose Setup Simulation... . Change the end time to 10 ms. Run the simulation, with the simulated scope set to 50 μs/div. Show the simulated waveforms on Plot 14-5. Label the axes of the plot.

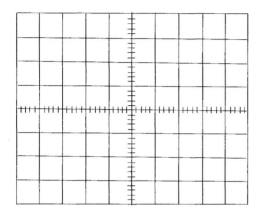   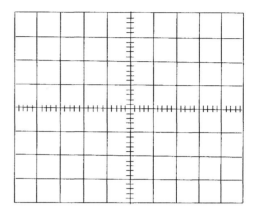

Plot 14-5 Simulated limiter waveform.   **Plot 14-6** Actual limiter waveform.

14. Download the configuration to the PAM. Use your oscilloscope to observe the actual waveforms at OUT1 and OUT2. Sketch your results in Plot 14-6. You may observe a small difference from the simulation. Record your observations:

Observations:_____

_____

Comparator Application – Tank Experiment
15. The remaining steps in this Part will simulate the pressure sensor and amplifiers in the tank experiment described in the text for the Programmable Analog Design feature. The input represents the amplified signal from a pressure sensor; the output represents the signal to turn on and off the fill pump. Open a new instance of AD2. Place a comparator CAM in the center of the design window. In the *Set CAM Parameters* window, set the hysteresis for 40 mV and accept other defaults. Connect InputCell1 to the input of the comparator and connect the output of the comparator to OutputCell1. The configuration is shown in Figure 14-15.

16. In the AD2 design window, place a function generator on InputCell1 and scope probes before and after the comparator. Set the simulated function generator for a 100 mV peak amplitude (200 mV$_{pp}$) sine wave at 5.0 kHz sine wave with default values for offsets. In the *Simulate* menu, choose Setup Simulation and choose a *Start Time* of 0 (default) and an *End Time* of 500 μs (500 u). Run the simulation. Set the scope Time per Division to 50 μs. Sketch the result of the simulation in Plot 14-7, adding labels to both axes.

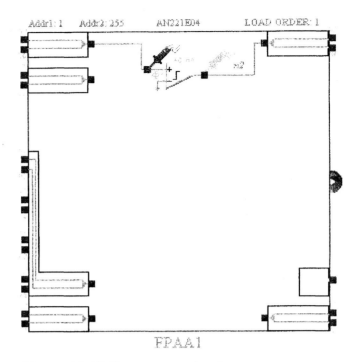

FPAA1

Figure 14-15 Comparator circuit.

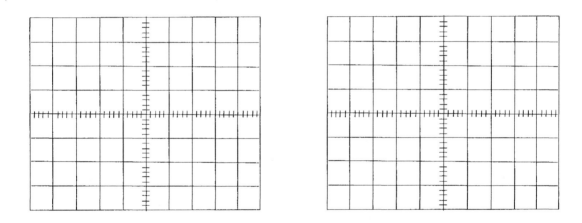

Plot 14-7 Simulated comparator waveform.   **Plot 14-8** Actual comparator waveform.

17. Download the configuration to the PAM. Connect the function generator to Input1P at
     the edge of the PAM and set it for a 200 mV$_{pp}$ 5.0 kHz sine wave, with no offset.
     Observe the generator input and the comparator output at Out1. Sketch the signals you
     see on Plot 14-8.

     Observations:_____

Conclusion: Part 2

Questions: Part 2

1. In some applications a zero-crossing detector that indicates when the input signal crosses the reference voltage in *both* directions is desired.
   (a) How would you develop an ASP configuration to do this?

   (b) Where might a circuit such as this be useful?

2. Under what conditions can a peak detector function as an unfiltered half-wave rectifier?

3. (a) How does a longer decay time constant affect the output of the peak detector?

   (b) How does a higher frequency affect the output of the peak detector?

4. How could you use the limiter CAM and gain stage to create a signal that closely resembles a square wave?

5. In step 15, hysteresis was specified for the comparator in the tank application. What do you think will happen in the actual implementation if there were no hysteresis?

Experiment 15-A Active Filters

A filter is a circuit that produces a prescribed frequency response. Active filters contain resistors, capacitors, and sometimes inductors as well as an active element such as an operational amplifier. Active filters can achieve frequency response characteristics that are nearly ideal for reasonable cost for frequencies up to about 100 kHz. Above this, active filters are limited by bandwidth.

The Butterworth form of filter has the flattest passband characteristic. Since a flat passband is generally the most important characteristic, it will be investigated in Part 1 of this experiment. The following page illustrates the method you will use for specifying the components of a filter. The method is entitled *Design Guidelines for a Butterworth Filter* and includes an example. The filter has the same characteristics as that used in the Application Activity, but the frequency is lower than in the AA to allow you to use the standard LM741C op-amp.

In Part 2, a state-variable filter is tested. State-variable filters can be designed for high Q band-pass filters ($Q > 100$) and they can be tuned to different frequencies. The bandwidth and Q can even be adjusted after the filter has been constructed. Because of the extra complexity, state-variable filters are generally used as band-pass filters, but they also have separate low-pass and high-pass outputs available from the same filter (band reject is also possible with an extra op-amp). State-variable filters are available in IC packages that allow the user to control gain, bandwidth, and Q using only external resistors.

Reading
Floyd, *Electronic Devices*, Eighth Edition, Chapter 15

Key Objectives
Part 1: Build and test a Butterworth low-pass active filter for a specific frequency and order.

Part 2: Construct a state-variable band-pass filter, measure and plot its frequency response. Compute and measure the center frequency and Q of the filter.

Components Needed
Part 1: Four-pole Low-Pass Filter
Resistors: one 1.5 kΩ, four 8.2 kΩ, two 10 kΩ, one 22 kΩ, one 27 kΩ
Capacitors: four 0.01 μF, four 1.0 μF
Two LM741C op-amps

Part 2: State-Variable Filter
Resistors: three 1.0 kΩ, three 10 kΩ, one 100 kΩ
Capacitors: two 0.1 μF, two 1.0 μF
Three LM741C op-amps

Design Guidelines for a Butterworth Filter

You can design your own Butterworth low-pass or high-pass active filter by using the following guidelines:

(1) Determine the number of poles necessary based on the required roll-off rate. Choose an even number, as an odd number will require the same number of op-amps as the next highest even number. For example, if the required roll-off is –40 dB/decade, specify a two-pole filter.

(2) Choose R and C values for the desired cutoff frequency (R_A, R_B, C_A and C_B). For best results, choose resistors between 1 kΩ and 100 kΩ. The values chosen should satisfy the cutoff frequency as given by the equation:

$$f_c = \frac{1}{2\pi RC}$$

(3) Choose resistors R_f and R_i that give the gains for each section according to the values listed in Table 15-1. The gain is controlled only by R_f and R_i. Solving the closed-loop gain of a noninverting amplifier gives the equation for R_f in terms of R_i:

$$R_f = (A_v - 1) R_i$$

Table 15-1
Butterworth Low-Pass and High-Pass Filters

Poles	Gain Required		
	Section 1	Section 2	Section 3
2	1.586		
4	1.152	2.235	
6	1.068	1.586	2.483

Example: A low-pass Butterworth filter with a roll-off of approximately –80 dB/decade and a cutoff frequency of 2.0 kHz is required. Specify the components.

Step 1: Determine the number of poles required. Since the design requirement is for a roll-off rate of approximately –80 dB/decade, a four-pole (two-section) filter is required.

Step 2: Choose R and C. Try C as 0.01 µF and compute R. Computed $R = 7.96$ kΩ. Since the nearest standard value is 8.2 kΩ, choose $C = 0.01$ µF and $R = 8.2$ kΩ.

Step 3: Determine the gain required for each section and specify R_f and R_i. From Table 15-1, the gain of section 1 is required to be 1.152 and the gain of section 2 is required to be 2.235. Choose resistors that will give these gains for a noninverting amplifier. The choices are determined by again considering standard values and are shown on the completed schematic in Figure 15-1.

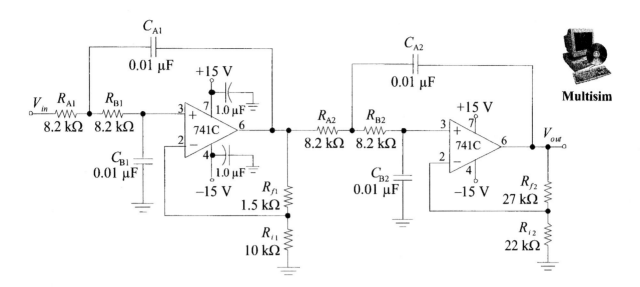

Figure 15-1

Application Activity

Part 1: Four-pole Low-Pass Filter

1. Measure and record the components listed in Table 15-2. If you cannot measure the capacitors, show the listed value.

Table 15-2

Component	Listed Value	Measured Values			
		A1	B1	A2	B2
R_{A1}, R_{B1}, R_{A2}, R_{B2}	8.2 kΩ				
C_{A1}, C_{B1}, C_{A2}, C_{B2}	0.01 μF				
R_{i1}	10 kΩ				
R_{f1}	1.5 kΩ				
R_{i2}	22 kΩ				
R_{f2}	27 kΩ				

2. Construct the four-pole low-pass active filter shown in Figure 15-1. Install a 10 kΩ load resistor. Connect a sine wave generator to the input. Set it for a 500 Hz sine wave at 1.0 V rms. The voltage should be measured at the generator with the circuit connected. Set the voltage with a voltmeter and check both voltage and frequency with the oscilloscope. Measure V_{RL} at a frequency of 500 Hz, and record it in Table 15-3.

3. Change the frequency of the generator to 1000 Hz. Readjust the generator's amplitude to 1.0 V rms. Measure V_{RL}, entering the data in Table 15-3. Continue in this manner for each frequency listed in Table 15-3.

185

4. Graph the load voltage (V_{RL}) as a function of frequency on Plot 15-1.

Table 15-3

Frequency	V_{RL}
500 Hz	
1000 Hz	
1500 Hz	
2000 Hz	
3000 Hz	
4000 Hz	
8000 Hz	

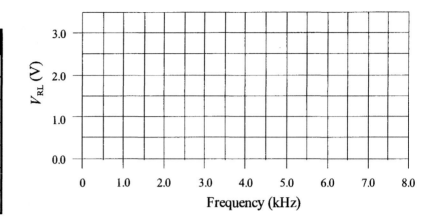

Plot 15-1

5. A Bode plot is a log-log plot of voltage versus frequency. It allows you to examine the data over a larger range than is possible with linear plots. Replot the data from the filter onto the log-log plot shown in Plot 15-2.

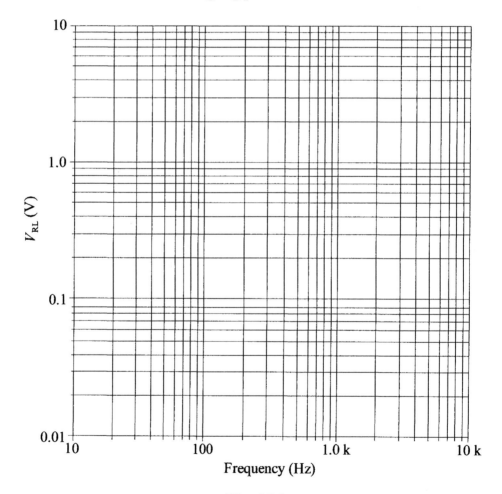

Plot 15-2

186

Conclusion: Part 1

Questions: Part 1

1.   (a) From the frequency response curves, determine the cutoff frequency for the filter in this experiment.

    (b) Compute the average R and C for your active filter (Table 15-1). Use the average values of each to compute the cutoff frequency.

2.   (a) What is the measured voltage gain of active filter in the passband?

    (b) What should it be?

3.   Using the Bode plot, predict V_{out} at a frequency of 20 kHz.

4.   The theoretical roll-off for your filter is –80 dB/decade. How does your actual filter compare to this theoretical roll-off rate?

5.   (a) Using the measured values of R_{i1} and R_{f1}, compute the actual gain of the first section. Compare this to the required gain in Table 15-1.

    (b) Repeat for the second section using R_{i2} and R_{f2}.

187

Part 2: State-Variable Filter

1. Measure and record the values of the resistors and capacitors listed in Table 15-4.

Table 15-4

Component	Listed Value	Measured Value
R_1	10 kΩ	
R_2	10 kΩ	
R_3	10 kΩ	
R_4	1.0 kΩ	
R_5	100 kΩ	
R_6	1.0 kΩ	
R_7	1.0 kΩ	
C_1	0.1 μF	
C_2	0.1 μF	

Table 15-5

Quantity	Computed	Measured
Center frequency, $f_0 =$		
$V_{pp(center)} =$		
Upper cutoff, $f_{cu} =$		
Lower cutoff, $f_{cl} =$		
Bandwidth, $BW =$		
$Q =$		

2. For the state-variable filter shown in Figure 15-2, compute the center frequency, f_0, and the Q of the circuit. For reference, the equations for calculating these parameters are as follows:

    The center frequency is:

$$f_0 = \frac{1}{2\pi R_4 C_1} = \frac{1}{2\pi R_7 C_2}$$

    The Q is determined by the gain of the first integrator as follows:

$$Q = \frac{1}{3}\left(\frac{R_5}{R_6} + 1\right)$$

    Compute the bandwidth, BW, by dividing f_0 by the Q. Enter these computed values in Table 15-5.

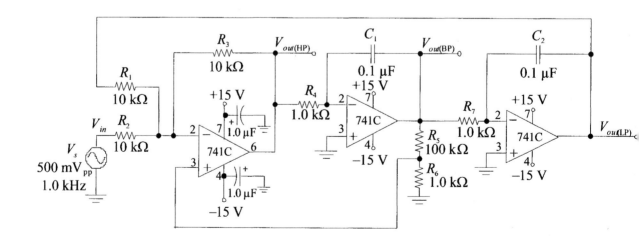

Figure 15-2

3. Construct the circuit shown in Figure 15-2. You will need the 1.0 µF bypass capacitors on the power supplies for only one op-amp, as shown. Set the function generator for a 500 mV$_{pp}$ sine wave at 1.0 kHz. Test the response from the band-pass output (center op-amp) by slowly increasing the frequency of the function generator as you observe the response. You should observe a dramatic peak in the output at the center frequency.

4. Observe the output with the oscilloscope and tune the center frequency, f_0 (maximum output). Measure this frequency and the peak-to-peak output voltage, $V_{pp(center)}$. Then carefully vary the frequency above and below the center until the output drops to 70.7% of the maximum voltage. These are the upper and lower cutoff frequencies, f_{cu} and f_{cl}, for the filter. The measured bandwidth is the difference between f_{cl} and f_{cu}. Record the measured values in Table 15-5. If available, you can obtain better accuracy if you use a frequency counter for the frequency measurements.

5. To obtain a better idea of the frequency response of your filter, measure and record the peak-to-peak output voltage as a function of frequency for the values listed in Table 15-6. Plot the response in Plot 15-3. Include the voltage for the center frequency on your plot. Because of the large dynamic range of the data, the plot is logarithmic.

Table 15-6

Frequency	Output voltage, V_{pp}
100 Hz	
200 Hz	
500 Hz	
1.0 kHz	
1.5 kHz	
2.0 kHz	
2.5 kHz	
3.0 kHz	
4.0 kHz	
5.0 kHz	
10 kHz	
20 kHz	

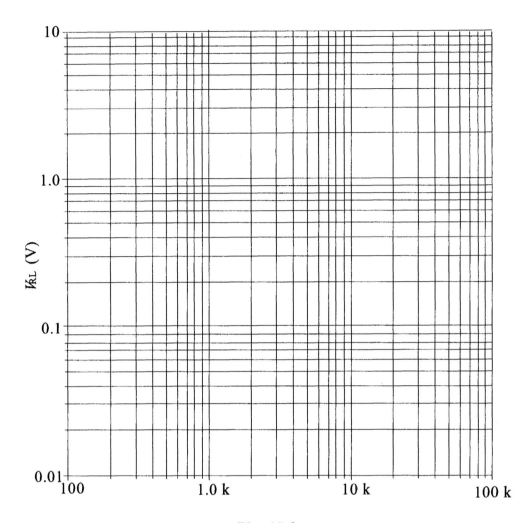

Plot 15-3

6. In addition to the band-pass output, the state-variable filter has low-pass and high-pass outputs as shown on Figure 15-2. Test these outputs in the circuit you constructed. What is the gain in the passband? Notice the gain change near the cutoff frequency. Can you think of a way of eliminating the peaking that occurs?

Observations:_____

_____

Conclusion: Part 2

Questions: Part 2

1.   The input signal for the filter was set for only 500 mV$_{pp}$. Under what circumstances could a larger signal be used without driving the output into saturation?

2.   Give the principal advantages of the state-variable filter in this experiment over the VCVS filter in Part 1.

3.   (a) What change would you make to the circuit in this experiment to raise the center frequency by a factor of two?

       (b) What change would you make to lower the Q of the circuit?

4.   (a) Assume the inputs of a summing amplifier are connected to the low-pass and high-pass outputs of the state-variable filter in this experiment. What type of filter does this form?

       (b) Name an application for this type of filter.

Multisim Simulation

Multisim

Multisim files for the lab manual are on the website www.prenhall.com/floyd. Open the Multisim file Experiment_15_Butterworth-nf. The circuit is the low-pass filter shown in Figure 15-1. Set up the Bode Plotter to show the response of the filter. Measure the midband gain and the cutoff frequency. Compare the simulation with your experimental result.

Experiment 15-B Programmable Analog Design

A filter is a circuit that passes certain frequencies and attenuates or rejects all other frequencies. Stated in other words, a filter is a circuit that produces a prescribed frequency response.

An active filter contains resistors, capacitors, and inductors plus one or more amplifier stages to achieve a desired frequency response. The Anadigm ASP on the PAM is easily programmed for various types of filters that are usable up to about 400 kHz. Before filter design software tools like AD2 became available, it took many hours of tedious work with a calculator and graph paper to design a filter. The AD2 software is easy to use and allows you to design any of the active filters described in the textbook in just seconds. This includes Butterworth, Chebyshev, and Bessel filters. In addition, AD2 can be used to design the Inverse Chebyshev and Elliptic filters that are not covered in the textbook.

There are two methods to design a filter in AD2:

- Select and place one or more filter CAMs and then set the desired parameters. The bilinear filter (*FilterBilinear*) CAM uses a single-pole like the simple *RC* filters in the textbook. The biquadratic filter (*FilterBiquad*) CAM uses a pair of poles plus it allows you to set the *Q* value independently to achieve the Butterworth, Chebyshev or Bessel filter response. You will use this method for Part 1 to make a single-pole low-pass filter.

- Use the AnadigmFilter design tool, which is a sub-program within AD2. AnadigmFilter is a graphical interface that allows you to select the filter type and response and then enter the corner frequency, passband, attenuation and desired passband ripple as values. Alternately, you can just draw a picture of the filter shape (the frequency response) and it will calculate the parameters for you using its "Automatic" filter model. When finished specifying the filter, just click the *To AnadigmDesigner2* tab and AnadigmFilter builds the filter for you in the AD2 design window. This method will be used in Parts 2 and 3 of this experiment.

In Part 2, you will construct a low-pass filter (this part can be done as a tutorial exercise up through step 13). In Part 3, you will construct a band-pass filter. A band-pass filter is a combination of a high-pass filter and a low-pass filter that together define a central frequency for which signals are passed through. You will build such a filter and then explore its features.

Reading

Floyd, *Electronic Devices*, Eighth Edition, Chapter 15, including review of the section on Programmable Analog Design.

Key Objectives

Part 1: Configure and test a single-pole low-pass filter using a bilinear filter.

Part 2: Configure and test a single-pole low-pass filter using AnadigmFilter and then measure and plot the frequency response.

Part 3: Configure and test a Chebyshev band-pass filter with specific characteristics using the AnadigmFilter tool.

Components Needed

Part 1: Single-Pole Low-Pass Filter Using the Bilinear Filter CAM
Programmable Analog Module
Function generator with BNC to grabber tip connectors. (A sweep generator is optional, but can be used to show the response.)

Part 2: Single-Pole Low-Pass Filter Using AnadigmFilter
Programmable Analog Module
Function generator with BNC to grabber tip connectors

Part3: Chebyshev Band-Pass Filter Using AnadigmFilter
Programmable Analog Module
Function generator with BNC to grabber tip connectors

Part 1: Single-Pole Low-Pass Filter using the Bilinear Filter CAM

1. Open a new instance of AD2. Place a bilinear filter (*FilterBilinear*) CAM in the upper center of the design window. In the *Set CAM Parameters* window, choose the default parameters except choose the Non-Inverting option and set the Corner Frequency to 5 kHz. (This will create a low-pass filter with a gain of 1.) Wire InputCell1 to the input of the filter and the output of the filter to OutputCell1. Add simulation scope probes to the input and output of the filter CAM as shown in Figure 15-3.

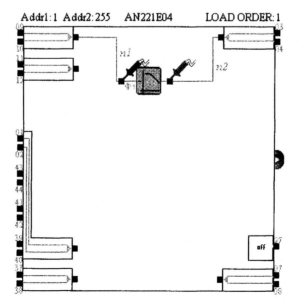

Figure 15-3 Single-pole low-pass filter using a bilinear filter CAM.

2.   In the *Simulate* menu, choose Setup Simulation… and change the End Time to 2 ms. Open the simulated signal generator using the shortcut G key (or use the *Simulate* menu and choose Create Signal Generator…). Connect the signal generator to InputCell1. Select the Sine Wave option and set the Peak Amplitude for 1.5 V (this is 3.0 V_{pp}) and the Frequency for 500 Hz (1/10 of the filter corner frequency).

3.   Run the simulation. Set the simulated oscilloscope Time Per Division to 200 µs to show one complete cycle of the two waveforms. Sketch the results in Plot 15-4. Label the axes.

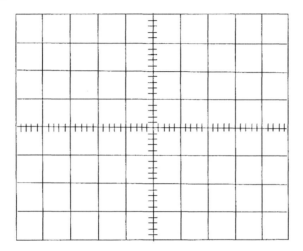

   Plot 15-4 Input and output waveforms from a single-pole low-pass filter.

4.   Select the *Cursor* button on the simulated oscilloscope to show the exact time and voltage for each waveform in the simulation. For the 500 Hz case, use the cursor to read the time from the zero-crossing of the input to the zero-crossing for the output of the filter. You can get a more accurate reading by spreading the waves out on the display. Record the time difference in these signals as Δt in the third column. Calculate the phase shift by dividing Δt by T (the period) and multiplying by 360°. The phase shift for output signals occurring later in time than the input (delayed in time, shifted to the right) is shown with a minus sign. Check that the input peak voltage (1.5 V) is displayed correctly and measure the peak voltage at the output of the filter. Record this data on the first line of Table 15-7.

Table 15-7 Simulated input and output voltage and time.

f	T	Δt	Phase shift	Input Peak Voltage, V_p	Output Peak Voltage, V_p
500 Hz	2000 µs				
2.5 kHz	400 µs				
5.0 kHz	200 µs				
10 kHz	100 µs				
50 kHz	20 µs				

5. Change the frequency of the signal source to 2.5 kHz. Repeat the simulation. Use the slider beneath the screen to display the last full cycle, to allow the simulation to settle. Repeat the measurements in step 4 for the 2.5 kHz signal, again measuring Δt at the zero crossing points. (You can read Δt more accurately by changing the Time Per Division to 10 µs for this particular reading.)

6. Repeat the time and voltage measurements for the remaining frequencies shown in Table 15-7, choosing appropriate settings of the scope Time Per Division and Volts Per Division controls to show a full cycle (or less for the time measurements). You should make readings near the end of the data to allow the simulation to settle.

7. Connect your function generator to Input1P at the edge of the PAM and set it for a 6.0 V_{pp}, 500 Hz sine wave, with no offset. (Figure 13-10 shows the Input1P.) Download the configuration file to the PAM. The output signal is attenuated a little bit by the filter and by a factor of two in the conversion process from differential to single-ended so the output should have a peak voltage of a little less than 1.5 V at the 500 Hz frequency (see Figure 14-11 for the idea). The time measurements should be made at the zero-crossing with the signals centered vertically on the scope. It is more accurate to use cursors if they are available on your scope. Record the measured data in the first line of Table 15-8.

Table 15-8 Measured input and output voltage and time.

f	T	Δt	Phase shift	Input Peak Voltage, V_p	Output Peak Voltage, V_p
500 Hz	2000 µs				
2.5 kHz	400 µs				
5.0 kHz	200 µs				
10 kHz	100 µs				
50 kHz	20 µs				

8. Repeat step 7 for each frequency in Table 15-8. Each time you change the generator, check that the waveforms are centered. You may need to adjust the amplitude of your function generator slightly to ensure it stays at 6.0 V_{pp} for each measurement. Record measured data in Table 15-8.

9. Compare the simulated filter response with the actual filter response.

Observations:

_____

_____

Conclusion: Part 1

Questions: Part 1

1. In step 4, you were instructed to measure time differences at the zero crossing point of the signals. Why is this more accurate than measuring time differences at the peaks of the signals?

2. Based on your experience with this single-pole low-pass filter, do you think that the output amplitude is smaller and is delayed in phase for all single-pole low-pass filters when the frequency increases?   Explain your answer.

3. Compare the measured response of the ASP filter to the ideal response of a one-pole filter given in Figure 15-1 of the textbook. In particular, look at the response at $0.1f_c$, f_c, and $10f_c$.

Part 2: Single-Pole Low-Pass Filter Using AnadigmFilter

1. In this step, you will start with the same filter that was used in Part 1. If you have that circuit in the AD2 design window, proceed to step 2. Otherwise, open a new instance of AD2. Place a bilinear filter (*FilterBilinear*) CAM in the upper center of the design window. In the *Set CAM Parameters* window, choose the default parameters except choose the Non-Inverting option and set the Corner Frequency to 5 kHz. (This will create a low-pass filter with a gain of 1.) Wire InputCell1 to the input of the filter and the output of the filter to OutputCell1. Add simulation scope probes to the input and output of the filter CAM as shown in Figure 15-3 (in Part 1).

2. You will now create and add another bilinear filter to the AD2 display window. In the *Tools* menu, select AnadigmFilter to start this tool. Point to and select the Full Screen option in the upper right corner. Select View and make sure the PoleZero Plot option is not checked and that the Tool Bar and Status Bar options are checked. You will see a display like Figure 15-4. Take time to study each of the following elements:

- The Frequency Response (Hz) plot. Notice that the Magnitude, which represents the gain, is in decibels and the frequency axis is logarithmic. This plot is equivalent to Figure 15-1 in the textbook
- Filter Type, selected for Low-pass option.
- Filter Approximation, selected for Butterworth.
- A list of the different filter CAMs (called "Modules" in AnadigmFilter) that will be used in series to construct the finished filter.
- Parameter settings:
  - Pass Band Ripple
  - Pass Band Gain
  - Stop Band Attenuation
  - Pass Band Freq.
  - Stop Band Freq.
  - Target

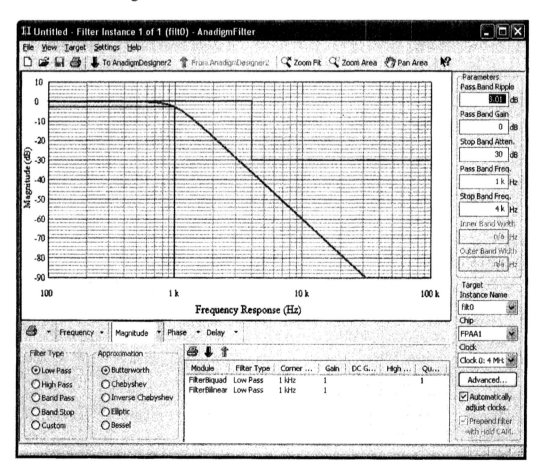

Figure 15-4 AnadigmFilter design window.

3.   The Frequency Response plot shows three lines:
   - The Stop Band Frequency line is red.
   - The Pass Band Frequency line is blue.
   - The Frequency Response line is green and always lies within the bounds set by the upper red Stop Band line and the lower blue Pass Band line.

   Carefully point to the vertical part of the Stop Band Frequency (red) and drag it slowly to the right to 10 kHz. Notice that several things change as you move it:
   - The roll-off (down hill) portion of the Frequency Response became less steep at about 6 kHz.
   - The bilinear filter (*FilterBilinear*) disappeared from the list of filter modules, leaving only the biquad filter (*FilterBiquad*) (a two-pole filter).

4.   Now drag the vertical part of the Stop Band Frequency further to the right to 100 kHz. As you drag it, notice that:
   - The roll-off (down hill) portion of the Frequency Response changed to a less steep slope as the Stop Band moved past about 37 kHz.
   - A bilinear filter (with only one-pole) replaced the biquad filter.
   - Another decade of frequency appeared on the *x*-axis.

5.   Point to the vertical part of the Pass Band Frequency (blue) and slowly pull it to the right so that its value is very close to 5 kHz. Notice that several things change as you move it:
   - The Pass Band Freq. window background turned light green to let you know that you are changing this value with your cursor.
   - The green Frequency Response line moved with the corner of the Pass Band Frequency.
   - When the Frequency Response line touched the inside corner of the red Stop Band line, the line suddenly got steeper again and the bilinear module was replaced by the biquad module.
   - The *Corner Frequency* entry associated with the filter module tracked the changes in the light green Pass Band Freq. window.

6.   Point to the Pass Band Freq window, highlight the current entry and then type in a new value = 5000 (The software will not accept the "k" entry for thousands).

7.   Drag the right hand horizontal portion of the red Stop Band Frequency up until the Frequency Response snaps up again and the bilinear module appears in place of the biquad entry, indicating a single-pole response. Notice that the Stop Band Atten. window is now light green and tracking the movement of the Stop Band line.

   At this point you should have a filter with the following settings:
   - Filter Type = Low-pass
   - Approximation = Butterworth
   - Filter Modules = only the bilinear filter

- Corner Frequency = 5 kHz.
- Gain = 1
- No entry for Quality Factor (does not apply to single-pole filters)
- Passband Ripple = 3.01 dB
- Passband Gain = 0 dB
- Stop Band Attenuation = 23 dB or smaller
- Pass Band Freq. = 5 kHz
- Stop Band Freq. = 100 kHz (approximately)

8.   Now point and click on the tab labeled *To AnadigmDesigner2* to build this filter in your AD2 design window. Wire a copy of the signal from InputCell1 to the input of this new CAM and wire the output to OutputCell2. Add a simulation scope probe to the output of the new filter. Connect a simulated signal generator to InputCell1, setting it for a 5 kHz sine wave with an amplitude of 1.5 V. Your design window should look like Figure 15-5.

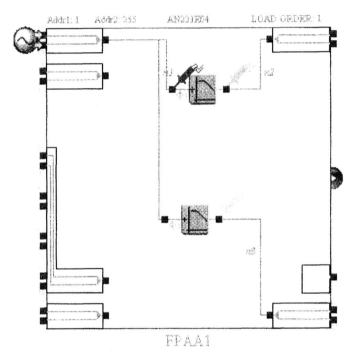

Figure 15-5 Low-pass filters from bilinear CAM and from AnadigmFilter.

9.   Point to each filter CAM in turn to see the parameter settings. If they are not exactly the same, then go back to the original sources, AnadigmFilter and the bilinear CAMs, and correct the settings to be the same. The two filters were designed by different methods, but the end result is the same. Simulate the circuit and observe the results.

Observations:

_____

_____

10. Start a new design in AD2 (File > New, Save changes = No). Close the previous instance of AnadigmFilter and then restart it (*Tools* > AnadigmFilter) to associate a new filter design with the new AD2 design. Set up a multipole Butterworth low-pass filter. Move the Stop Band vertical line to 10 kHz. Move the Pass Band vertical line to 5 kHz (enter exact 5000 Hz and 10000 Hz values after moving the stop band lines). The Ripple, Pass Band Gain, and Stop Band Atten. should be at their default values (3.01 dB, 0 dB, 30 dB, respectively). You should see two biquad filters and one bilinear filter (5 poles total) in the module list. Click the tab labeled *To AnadigmDesigner2*, and the new design will be implemented in AD2.

11. Wire InputCell1 to the input of the filter and wire the output of the filter to OutputCell1. Place simulated scope probes on the input and output of the filter. Connect the simulated signal generator to InputCell1, setting it for a 500 Hz sine wave at 1.0 V amplitude. Your design window should look like Figure 15-6.

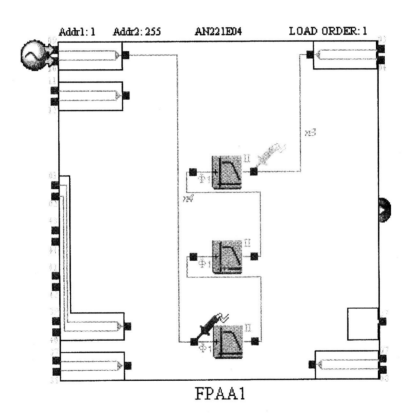

Figure 15-6 Five-pole low pass filter.

12. In the *Simulate* menu, choose Setup Simulation... and change the End Time to 2 ms. Run the simulation. Set the simulated oscilloscope Time Per Division to 200 μs to show one complete cycle of the two waveforms. Measure the input peak voltage and the output peak voltage for the 500 Hz case and record the values in Table 15-9.

Table 15-9 Simulated input and output voltage
for a five-pole low-pass filter.

f	Input Peak Voltage, V_p	Output Peak Voltage, V_p
500 Hz		
2.5 kHz		
5.0 kHz		
10 kHz		
50 kHz		

13. Change the simulated signal generator frequency to 2.5 kHz and rerun the simulation. Record the input and output peak voltages in Table 15-9. Continue like this for the remaining frequencies. Generally, measurements should be made near the end of the simulation to allow it to settle. Notice how the response "drops like a rock" due to the five-pole filter. The last value (50 kHz) will likely be too small to make an accurate reading using an oscilloscope on a real circuit.

14. Download the file to the PAM. Apply a 4 V_{pp} sine wave to Input1 (4 V_{pp} at the edge of the PAM is 2 V_{pp} at the input to the ASP, which is 1 V_p as in your simulation). If you have a sweep generator, you can set it up to show the frequency response on your oscilloscope. The generator sawtooth output will control the *x*-axis as described in the textbook in Section 15-7. If you do not have a sweep generator, change the input signal frequency slowly from well below the critical frequency to well above the critical frequency and notice how rapidly the output voltage rolls off above f_c. Measure the voltage at OUT1 when the input is 5.0 kHz (cutoff.)

V_{out} at 5.0 kHz = _____

15. A five-pole filter should have a roll-off rate of −100 dB per decade, which is −30 dB per octave (an octave is a factor of 2). One octave above the critical frequency is 10 kHz. Measure the voltage at OUT1 at 10 kHz and confirm that the filter is working as advertised.

V_{out} at 10 kHz = _____

Conclusion: Part 2

Questions: Part 2

1. In step 15, you made a simple test of the five-pole filter one octave above the critical frequency. Why is this a better choice than one decade above the critical frequency?

2. (a) In Plot 15-5, show the measured points for the five-pole low-pass filter that you tested in steps 14 and 15.

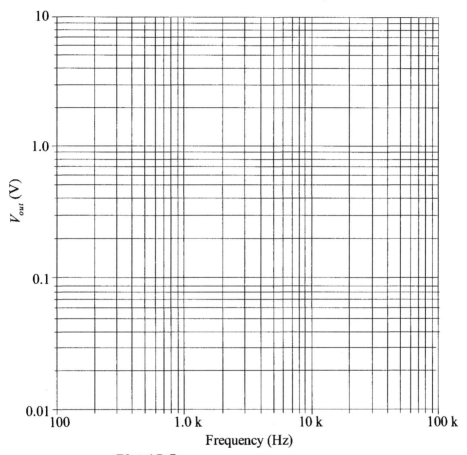

Plot 15-5

(b) Draw a straight approximation of the filter frequency response below f_c and a second straight line approximation of the filter frequency response above f_c. Use the result to predict the frequency at which the output will be 10 mV.

Predicted frequency for a 10 mV output = _____

Part 3: Chebyshev Band-Pass Filter using AnadigmFilter

1. Start a new instance of AD2. Close any previous instance of AnadigmFilter and then restart it for a new filter design. Configure a filter with these characteristics:

    *   Filter Type = Band Pass
    *   Approximation = Chebyshev (notice the ripples in the pass band region)
    *   Pass Band Ripple = 3.01 dB
    *   Pass Band Gain = 0 dB (same as 1 times gain)
    *   Stop Band Attenuation = 20 dB
    *   Center Frequency = 5 kHz
    *   Stop Band Width = 5 kHz (set this before setting the Pass Band Width)
    *   Pass Band Width = 2 kHz

    If you've entered all these values correctly, you should see two ripple peaks in the pass band and these filter modules in the area below the Frequency Response plot:

Module	Filter Type	Corner Freq	Gain	Quality Factor
• FilterBiquad	Band Pass	4.282 kHz	2.22	7.86
• FilterBiquad	Band Pass	5.839 kHz	2.22	7.86

    AnadigmFilter is proposing two filters in series to achieve this frequency response. The gains are both greater than 1 because each filter attenuates the signal to achieve the desired 1X gain at each peak. Quality Factor values (Q) are high, characteristic of a Chebyshev filter that has steep out-of-band roll-offs.

    The frequency response plot should look like Figure 15-7. If you see this response, then select the *To AnadigmDesigner2* tab to build this filter into your AD2 design window. Otherwise, review the settings for the filter and correct any errors.

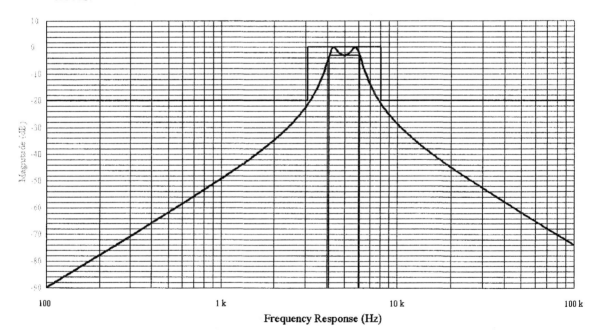

Figure 15-7 Frequency response plot for Chebyshev band-pass filter.

2. Continuing with the AnadigmFilter window, point your cursor anywhere in the
    Frequency Response Plot area. You will see a vertical dotted line that moves
    across the frequency plot with your mouse. Hold the mouse still a moment and a
    window appears that shows the exact frequency and the magnitude of the filter
    response at that frequency. Measure the frequencies of the peaks in the pass band:

    Peak #1 (left) at _____ kHz (You should see about 4.34 kHz)

    Peak #2 (right) at _____ kHz

3. Move the cursor to the point where the response curve (green) just touches the
    stop band (red). Measure the frequency at the ends of the stopband and confirm
    that the stopband is the specified 5 kHz:

    Edge of stopband (left) at _____ kHz

    Edge of stopband (right) at _____ kHz

4. Go to the AD2 window. You should see two new CAM icons wired in series; the
    icon art suggests a band-pass filter response. Wire InputCell1 to the input of the
    first filter CAM and the output of the second filter CAM to OutputCell1. Put the
    violet scope probe at the input and the green scope probe at the output. Install a
    signal generator at InputCell1 set for 1.5 Volts *Peak Amplitude* (3 V_{pp}) and 1 kHz.
    In the *Simulate* menu, click on Setup Simulation... ; set the simulation end time to
    10 ms. You should see an AD2 design window like Figure 15-8.

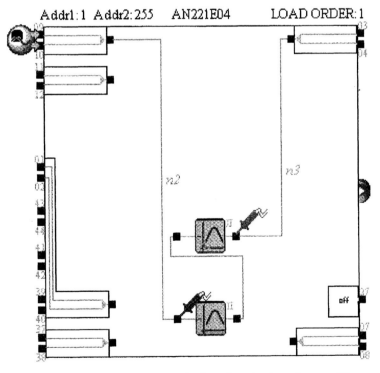

Figure 15-8 AD2 design for the band-pass filter.

5. Run the simulation to completion. On the oscilloscope, set the Time Per Div for 100 μs. Set Channel 2 Volts Per Div to 50 mV. Leave the time slider full to the left. You should see a display like Figure 15-9.

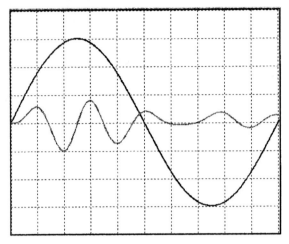

Figure 15-9 Start up response of 5 kHz center frequency band-pass filter.

Notice that the input starts by ringing. What is the ring frequency that you see?

$f_{ring} = $ _____

Why do you think the ring frequency is higher than the input signal? (*Hint*: The input has gone from a flat signal to a sine wave. In mathematical terms, the input is discontinuous.)
Reason for the higher initial ring frequency: _____

---

6. In general, you should observe simulations after steady-state performance is achieved, so drag the time slider to the right to show the last full cycle as in Figure 15-10. Since the applied signal is well outside the passband, the output signal amplitude is small.

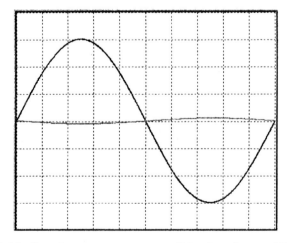

Figure 15-10 Steady-state response of the band-pass filter at 1 kHz.

7. Set the signal generator to the frequency of the first peak that you recorded in step 2 and run the simulation. In the Volts Per Division window, set Channel 1 and 2 to 500 mV. Set the Time Per Division to 50 μs. Push the time slider to the right to show the last two cycles as in Figure 15-11. Note that the output signal (green) is almost exactly the same amplitude as the input signal (violet) but is leading in phase. Because this is a characteristic of high-pass filters, you can conclude that the high-pass filter is dominant at this frequency and affects the response at frequencies below the passband.

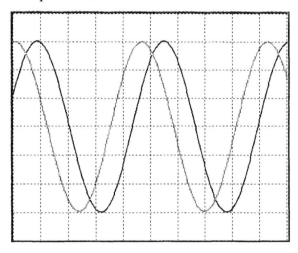

Figure 15-11 Response the band-pass filter at 4.34 kHz.

8. Set the signal generator to the frequency of the second peak that you recorded in step 2. Run the simulation again. Notice that the output signal (green) is now lagging the input. Low-pass filters have this characteristic, so you know that the low-pass filter is dominant at this frequency and affects the response at frequencies above the passband.

9. Download the filter to the PAM. Connect your function generator to INP1 (shown in Figure 13-10) and set it for a 3.0 kHz and 6.0 V_{pp} sine wave (1.5 V_p at the input to the ASP as in your simulation). Slowly move the frequency dial of your function generator from 3.0 kHz to 8.0 kHz. As you do so, watch the output signal at OUT1 on the oscilloscope.

Observations:

_____

_____

10. Using an oscilloscope, measure and record the peak output voltages at OUT1 for each frequency listed in Table 15-10. Check the function generator level each time you change frequencies to verify it remains at 6.0 V_{pp} (3.0 V_p) and record the peak value of the input.

Table 15-10 Output voltage for the band-pass filter.

f	Input Peak Voltage, V_p	Output Peak Voltage, V_p
3.1 kHz	3.0 V_p	
4.3 kHz		
5.0 kHz		
5.7 kHz		
8.2 kHz		

11. Plot the data from Table 15-10 in Plot 15-6.

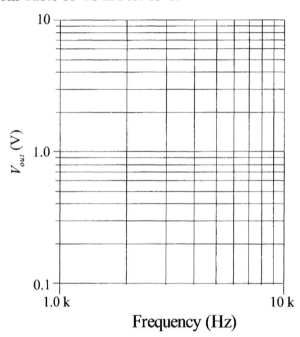

Plot 15-6 Frequency response of the Chebyshev band-pass filter

12. Compare your plotted data with the AnadigmFilter simulation. What similarities or differences do you see?

Observations:

_____

_____

Conclusion: Part 3

Questions: Part 3

1. AnadigmFilter designed two biquadratic filters in series to achieve the band-pass filter response. If the stop band attenuation had been specified as 30 dB instead of 20 dB, would more or less filters been required? Explain your answer.

2. (a) What evidence did you see in the experiment that the maximum overall gain of the band-pass filter is 1?

    (b) The maximum gain of the two biquadratic filters is 1, yet the gain of each individual filter was shown by AnadigmFilter as 2.22. Explain why.

3. The Chebyshev filter is much more popular than the Bessel filter, yet the Bessel filter has a smoother response curve. Why do you think this is true? (Try checking both responses on AnadigmFilter to find the answer.)

Experiment 16-A Oscillators

In electronic systems, there is almost always a requirement for one or more circuits that generate a continuous waveform. The output voltage can be a square wave, sine wave, sawtooth, or other periodic waveform. A free-running feedback oscillator is basically an amplifier that generates a continuous alternating voltage by feeding a portion of the output signal back to the input in the proper phase.

In Part 1, a Wien-bridge oscillator is investigated. This circuit is a popular *regenerative* oscillator, which is used to generate high quality sinusoidal waveforms. A portion of the output is fed back to the input in the proper amplitude and phase to reinforce the signal. This type of feedback is called *positive* feedback. For the standard Wien-bridge, the feedback network returns 1/3 of the output signal to the noninverting input. Therefore, the amplifier must provide a gain of 3 to overcome this attenuation and prevent oscillations from dying out. As explained in the text, the gain level needs to be controlled with automatic gain control (AGC) to maintain a good sine-wave. This control is accomplished by *negative* feedback. You will investigate an FET stabilized Wien bridge in this part.

In Part 2, two other types of regenerative oscillators are investigated. These are the Hartley and Colpitts types that differ mainly in the feedback network. The Hartley oscillator obtains feedback between two inductors in the tank circuit whereas the Colpitts oscillator obtains the feedback from between two capacitors in the tank circuit. You may want keep the Colpitts oscillator that you construct for an optional experiment at the end of Part 3, which is closely linked to the Application Activity in the text.

Part 3 introduces the versatile 555 timer chip, one of the first ICs of its type and still one of the most popular timers. Entire books of circuits have been written for this IC, and it retains its popularity for a variety of timing, measurement and control applications. You will investigate the basic astable multivibrator and specify the components for an astable circuit.

The optional investigation at the end of Part 3 combines the timer circuit in Part 3 with the Colpitts oscillator constructed in Part 2. You will need a switching transistor to complete the circuit. This circuit is equivalent to the modulated source given as the Application Activity in the text.

Reading
Floyd, *Electronic Devices*, Eighth Edition, Chapter 16

Key Objectives
Part 1: Construct and test an FET stabilized Wien-bridge oscillator.
Part 2: Construct and test the Colpitts and the Hartley oscillators and compare the computed and measured performance of both types.
Part 3: Calculate and measure parameters for an astable multivibrator.

Components Needed

Part 1: The Wien-Bridge Oscillator
Resistors: one 1.0 kΩ, three 10 kΩ
Capacitors: two 0.01 μF, three 1.0 μF
Two 1N914 signal diodes (or equivalent)
One 741C op-amp
One 2N5458 n-channel JFET transistor (or equivalent)
One 10 kΩ potentiometer

Part 2: The Hartley and Colpitts Oscillators
Resistors: one 1.0 kΩ, one 2.7 kΩ, one 3.3 kΩ, one 10 kΩ
Capacitors: two 1000 pF, one 0.01 μF, four 0.1 μF
Inductors: one 2 μH (use a length of #22 wire), one 25 μH, one 150 μH
One 2N3904 npn transistor (or equivalent)
One 100 Ω potentiometer

Part 3: The 555 Timer
Resistors: one 8.2 kΩ, one 10 kΩ, others specified by student
One 0.01 μF capacitor
One 555 timer

Optional Investigation – Modulated Source
Components from Part 2 and 3 plus:
One 1000 μF capacitor
One 1N914 diode
One J176 p-channel FET
Additional 10 kΩ resistor

Part 1: The Wien Bridge Oscillator

1. Measure R_1, R_2, C_1, and C_2 for this experiment. These components determine the frequency of the Wien bridge oscillator. Record the measured values in Table 16-1. If you cannot measure the capacitors, record the listed value.

2. Construct the basic Wien-bridge illustrated in Figure 16-1. Adjust R_f so that the circuit just oscillates. You will see that it is nearly impossible to obtain a clean sine wave as the control is too sensitive. If you have freeze spray available, try spraying it lightly on the components and observe the effect.

    Observations:_____

Table 16-1

Component	Listed Value	Measured Value
R_1	10 kΩ	
R_2	10 kΩ	
C_1	0.01 μF	
C_2	0.01 μF	

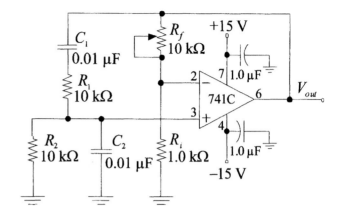

Figure 16-1

3. The bridge in step 2 has the problem of unstable gain and requires some form of automatic gain control (AGC) to work properly. Field-effect transistors are frequently used for AGC circuits because they can be used as voltage-controlled resistors for small applied voltages. The circuit illustrated in Figure 16-2 is an FET-stabilized Wien-bridge. Compute the expected frequency of oscillation from the equation:

$$f_r = \frac{1}{2\pi RC}$$

Use the <u>average</u> measured value of the resistance and capacitance listed in Table 16-1 to calculate f_r. Record the computed f_r in Table 16-2.

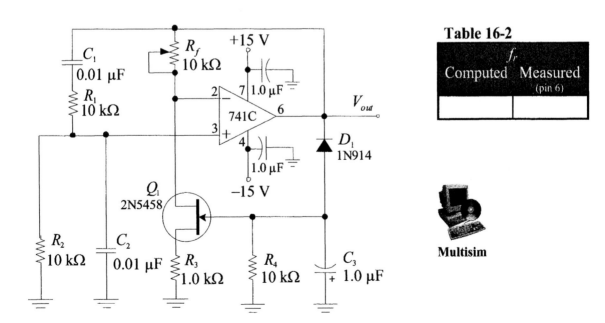

Figure 16-2

Table 16-2

f_r	
Computed	Measured (pin 6)

Multisim

213

4. Construct the FET-stabilized Wien-bridge shown in Figure 16-2. The diode causes negative peaks to charge C_3 and bias the FET. C_3 has a long time constant discharge path (through R_4) so the bias does not change rapidly. Note the polarity of C_3. Adjust R_f for a good sine wave output. Measure the frequency and record it in Table 16-2.

5. Measure the peak-to-peak output voltage, $V_{out(pp)}$. Then measure the peak-to-peak positive and negative feedback voltages, $V_{(+)(pp)}$ and $V_{(-)(pp)}$, and the dc voltage on the gate of the FET (V_G). Use two channels and observe the phase relationship of the waveforms. Record the voltages in Table 16-3.

Table 16-3

Measured Voltages			
$V_{out(pp)}$ (pin 6)	$V_{(+)(pp)}$ (pin 3)	$V_{(-)(pp)}$ (pin 2)	V_G

What is the phase shift from the output voltage to the positive feedback voltage?

_____

6. Try spraying freeze spray on various components while observing the output.

Observations: _____

7. Add a second diode in series with the first one between the output and the gate of the FET (See Figure 16-3). You may need to readjust R_f for a good sine wave. Measure the voltages as before and record in Table 16-4.

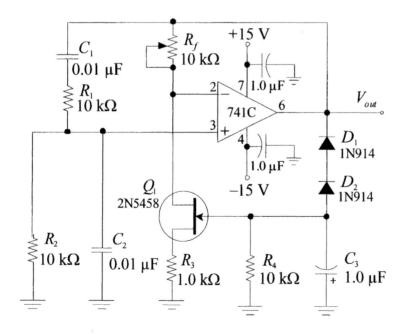

Figure 16-3

Table 16-4

Measured Voltages			
$V_{out(pp)}$ (pin 6)	$V_{(+)(pp)}$ (pin 3)	$V_{(-)(pp)}$ (pin 2)	V_G

Conclusion: Part 1

Questions: Part 1

1. In step 5, you measured the positive feedback voltage.

   (a) What fraction of the output voltage did you find?

   (b) Is this what you expect from theory?

2. Explain why adding a second diode in series with the first caused the output voltage to increase.

3. For the circuit in Figure 16-2, why is the positive side of C_3 shown on ground?

4. At what frequency would the Wien bridge of Figure 16-2 oscillate if R_1 and R_2 were doubled?

Part 2: The Hartley and Colpitts Oscillators

1. Measure and record the value of the resistors listed in Table 16-5.

2. Observe the class A amplifier shown in Figure 16-4. Using your measured resistor values, compute the dc parameters for the amplifier listed in Table 16-6. R_{E1} is a 100 Ω potentiometer that you should set to 50 Ω. Then construct the circuit and verify that your computed dc parameters are as expected. Record the measured values in Table 16-6.

Table 16-5

Resistor	Listed Value	Measured Value
R_1	10 kΩ	
R_2	3.3 kΩ	
R_{E1}	50 Ω *	
R_{E2}	1.0 kΩ	
R_C	2.7 kΩ	

* set potentiometer for 50 Ω

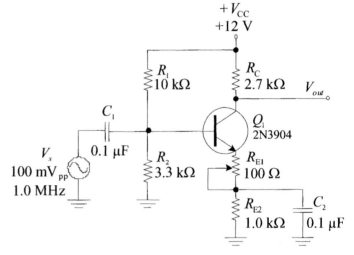

Figure 16-4

3. Calculate and record the computed ac parameters listed in Table 16-7. After you find r_e', the gain is calculated by dividing the collector resistance by the sum of the unbypassed emitter resistance and r_e'. (Assume that the potentiometer remains set to 50 Ω.) The ac voltage at the collector is calculated by multiplying the gain by the ac base voltage.

4. Set the function generator for a 100 mV$_{pp}$ signal at 1.0 MHz and measure the peak-to-peak voltage at the base and the collector. Use these measured values to determine the measured gain. Record the measured values in Table 16-7. The computed and measured values should agree within normal experimental uncertainty.

Table 16-6

DC Parameter	Computed Value	Measured Value
V_B		
V_E		
I_E		
V_C		

Table 16-7

AC Parameter	Computed Value	Measured Value
V_b		
r_e'		
A_v		
V_c		

216

5. Remove the signal generator and add the feedback network for a Hartley oscillator as shown in Figure 16-5. (L_2 can be wound by wrapping about 40 turns of #22 wire on a pencil). Adjust R_{E1} for the best sine wave. Compute the frequency of the Hartley oscillator and record the computed frequency in Table 16-8. Then, measure the frequency and the peak-to-peak voltage at the output and record them in Table 16-8.

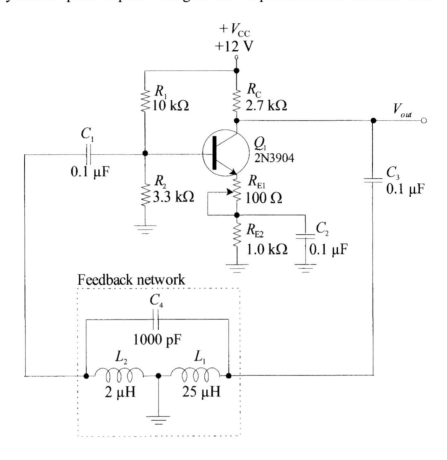

Figure 16-5

Table 16-8

Hartley Oscillator	Computed Value	Measured Value
Frequency		
Amplitude		

6. Observe what happens to the frequency and amplitude of the output signal when another 1000 pF capacitor is placed in parallel with C_4.

Observations: _____

_____

7. Replace the feedback network with the one shown in Figure 16-6. Adjust R_{E1} for a good sine wave output. This configuration is that of a Colpitts oscillator. Compute the frequency of the Colpitts oscillator, and record the computed frequency in Table 16-9. Then, measure the frequency and the peak-to-peak voltage at the output and record them in Table 16-9.

Feedback network

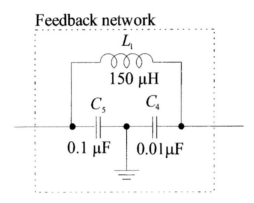

Table 16-9

Colpitts Oscillator	Computed Value	Measured Value
Frequency		
Amplitude		

Figure 16-6

Note: If you construct the optional modulated source described after Part 3, retain the Colpitts oscillator on your protoboard.

Conclusion: Part 2

Questions: Part 2

1. In step 6, you observed a change in the amplitude of the output signal when a capacitor was placed in parallel with C_4. Since the gain of the class A amplifier remained the same, what conclusion can you draw about the effect of the change on the amount of feedback?

2.   What are the two conditions required for oscillation to occur in an *LC* oscillator?

3.   Summarize the key difference between a Colpitts and a Hartley oscillator.

Part 3: The 555 Timer

1.   Measure and record the resistances of R_1 and R_2 and the capacitance of the timing capacitor C_{ext} with values shown in Table 16-10. If you cannot measure the capacitor, enter the listed value in the measured column.

Table 16-10

Component	Listed Value	Measured Value
R_1	8.2 kΩ	
R_2	10 kΩ	
C_{ext}	0.01 μF	

2.   One of the requirements for most digital circuits is a clock, a series of pulses used to synchronize the various circuit elements of a digital system. In the astable mode, a 555 timer can serve as a clock generator.
        An astable circuit using the 555 timer is shown in Figure 16-7. There are two timing resistors, R_1 and R_2 and a timing capacitor, C_{ext}. The timing equations are given in the text and repeated here for reference.

$$f = \frac{1.44}{(R_1 + 2R_2)C_{ext}}$$

$$\text{Duty cycle} = \left(\frac{R_1 + R_2}{R_1 + 2R_2}\right)100\%$$

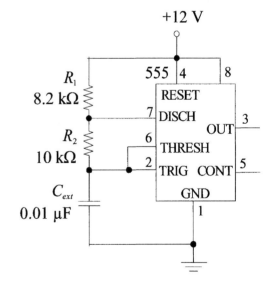

Figure 16-7

Calculate the expected timing parameters for the circuit shown in Figure 16-7. Enter the data in Table 16-11.

3. Construct the astable multivibrator circuit shown in Figure 16-7. Pin 5 (Control) is internally connected to a voltage divider. It can be left open for this experiment, although some people prefer to connect a small capacitor to ground to avoid noise.

Table 16-11

Component	Computed Value	Measured Value
Frequency		
Duty cycle		

Using an oscilloscope, measure the quantities listed in Table 16-11 and record the measured values.

4. With the oscilloscope, observe the waveforms across capacitor C_1 and the output waveform at the same time. On Plot 16-1, sketch the observed waveforms. Label the time and voltage on your plot and add a title.

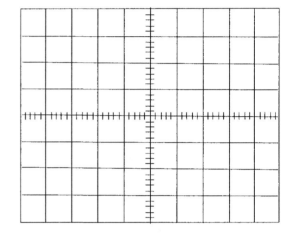

5. While observing the waveforms from Step 4, try placing a short across R_B. After viewing the result, remove the short.
Observations:

_____

_____

Plot 16-1

6. Specify the components for an astable multivibrator constructed with a 555 timer that has a frequency of 10 kHz. (The duty cycle is not critical). Draw your circuit in the space provided on the right.

7. Construct the circuit you specified in step 6, test it, and summarize your results.

Observations:

_____

Conclusion: Part 3

Questions: Part 3

1. How would you change the circuit in Figure 16-7 to maintain the same frequency but have a higher duty cycle?

2. If the power supply is increased to +15 V for the circuit in Figure 16-7, what change would you expect to see on the waveforms that you observed across C_{ext}?

3. In the basic astable circuit in Figure 16-7, it is not possible to obtain a 50% duty cycle. Explain why not.

4. (a) In step 4, you observed the voltage across the capacitor for the basic astable. circuit. How is the amplitude of this voltage related to the power supply voltage?

    (b) Does the power supply setting have an effect on the frequency from the 555 timer? Why or why not?

5. Look up the manufacturer's specification sheet for the 555 timer[1].
    (a) What is the maximum current the output can source or sink?

    (b) What happens to the output voltage when the timer supplies higher current?

Optional Investigation – Modulated Source

1. The Colpitts oscillator constructed in Part 2 can be combined with the timer constructed in Part 3 to form a modulated source very similar to the one in the Application Activity of the text. A large electrolytic capacitor (C_7) is useful to avoid noise. A diode (D_1) is added to the timer to force the duty cycle to be close to 50% and the p-channel FET (Q_2) is added to switch the output on and off. In addition, a 10 kΩ load resistor (R_L) is added to the circuit. Some components have been renumbered for the combined circuit. Construct the circuit in Figure 16-8.

---

[1] The specification sheet is available from Texas Instruments at http://www.ti.com

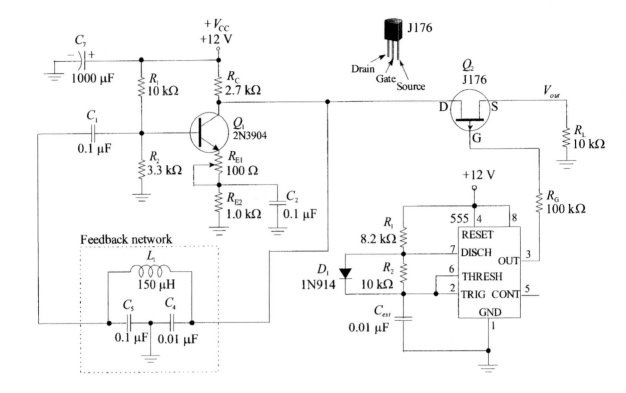

Figure 16-8

2. Observe the waveforms at the output of the oscillator, the timer, and across the load resistor. Because the oscillator and the timer are not synchronized, you will not be able to "freeze" both waveforms on the oscilloscope at the same time. Try changing the triggering channel as you look at two points in the circuit to see the effect. Summarize your observations in a short report.

Multisim Simulation

Multisim

Multisim files for the lab manual are on the website www.prenhall.com/floyd. There are two Multisim files for this experiment. The first is the Wien Bridge circuit from Part 1. Open the Multisim file Experiment_16_Wien Bridge-nf. The circuit is the same as the one in Figure 16-2 except a 2N5485 is used for the FET because the 2N5458 is not available in Multisim. Connect the scope and adjust the gain to start oscillations. Measure the frequency and compare the simulation to your experimental result.

The second file is Experiment_16_RFID source-nf, which simulates the circuit in Figure 16-8. You will need to set up the oscilloscope to view the various waveforms. Because the oscillator is not synchronous with the 555 timer output, you cannot "freeze" both the modulated output and the oscillator at the same time. This is a case where an external trigger can be useful. Try using the external trigger from the 555 output to stabilize the display on the modulated signal.

Experiment 16-B Programmable Analog Design

In most electronic systems, there is a requirement for one or more circuits that generate a periodic waveform. The waveform could be a square wave, sine wave, sawtooth, or some other periodic waveform. The textbook describes several different circuits that use positive feedback to make an oscillator. In Part 1 you will implement positive feedback in the form of a ring oscillator. The basic idea for a ring oscillator is to form an unstable loop by connecting an odd number of inverting gain stages in a loop that closes the oscillating signal on itself. The minimum number of stages that typically can be made to oscillate is three. Interestingly, ring oscillators illustrate a limitation of circuit simulators in general: they have a hard time dealing with positive feedback as you will see.

Ring oscillators are simple and easy to understand, but they don't achieve very accurate frequencies without some additional frequency-locking mechanism. They are widely applied to integrated circuit development work as a measure of the intrinsically achievable speed for each new IC process as it is developed. In this case, the oscillator frequency is a measure of the delay time through the circuit.

An entirely different form of oscillator can be called an "algorithmic oscillator". These oscillators are implemented in clock-driven systems like the Anadigm ASP where the amplitude of the next data point is determined by an algorithm that defines the whole periodic waveform point by point. Although the Anadigm ASP is clock driven, it is *not* a digital signal processor – the signals inside the Anadigm ASP really are analog samples and all the methods and mathematics of analog signal processing apply. The processing of these samples is done stepwise with the supplied clock.

In Part 2, you will explore an algorithmic oscillator that you have used in previous experiments – the sine wave oscillator. The functional details of this CAM were not seen because they are hidden inside the CAM definition. The output signals are very good sine waves when there are 50 or more analog samples processed for each cycle, but in this Part the useable limits will be explored to learn more about the ASP.

In Part 3, another algorithmic oscillator is introduced and tested. The arbitrary waveform generator (AWG) is a standard laboratory instrument that can be implemented in the ASP. An arbitrary waveform is created from values specified in a look-up table and converted to analog voltages by a digital-to-analog converter. The CAM name is *PeriodicWave*, which is a reference to the fact that it repeats the process of creating the waveform for each iteration through the lookup table.

Reading

Floyd, *Electronic Devices*, Eighth Edition, Chapter 16, including review of the section on Programmable Analog Design.

Key Objectives

Part 1: Configure and test several different ring oscillators.

Part 2: Configure and test the advanced features of the Anadigm sine wave oscillator CAM.

Part 3: Configure and test the arbitrary periodic waveform generator CAM.

Components Needed

Part 1: Ring Oscillators

Programmable Analog Module
Stereo Audio Cable with 3.5 mm male connectors each end,
approximately 3 feet length (supplied with each PAM)

Part 2: The Sine Wave Oscillator

Programmable Analog Module

Part 3: The Arbitrary Waveform Generator

Programmable Analog Module
Spreadsheet computer program (such as Microsoft Excel)

Part 1: Ring Oscillators

1. Close any open instances of AnadigmFilter and AD2 and start a new version of AD2. Place an inverting gain (*GainInv*) CAM in the design window. In the *Set CAM Parameters* window, set the gain to 10. Select this stage and use the *copy* and *paste* command to duplicate it two more times. Wire the three stages in series and connect the first input to InputCell1 and the final output to OutputCell1. Connect an external wire from OUT1P to IN1P and OUT1N to IN1N. This configuration forms a three-stage ring oscillator; it should look like the one shown in Figure 16-9.

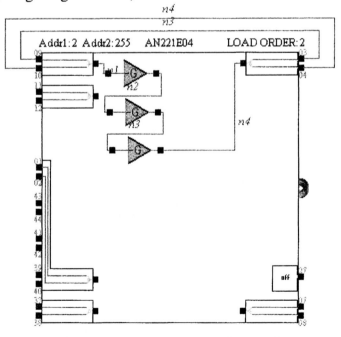

Figure 16-9 Three-stage ring oscillator.

2. Place a simulator scope probe at the output of the final *GainInv* CAM and start the simulator. You will see an error message that says: "This circuit does not have a voltage source. The simulation will be terminated." As mentioned in the introduction, circuit simulators frequently have trouble with positive feedback. This is because each stage is the voltage source for the following stage around the ring, so the simulator does not see a stand-alone source to work with. Even if the ring oscillator is completely inside the ASP, you would see the same error message.

3. Despite the fact that the simulator has trouble, you can implement the circuit by adding external feedback and see it work. On your PAM unit connect the OUTPUT jack to the INPUT jack using the stereo audio cable. Place your oscilloscope scope probe on OUT1. Download the AD2 configuration. You should see a sawtooth waveform; measure the frequency and amplitude and sketch the waveform in Plot 16-2. Label the axes.

Frequency _____ Amplitude _____

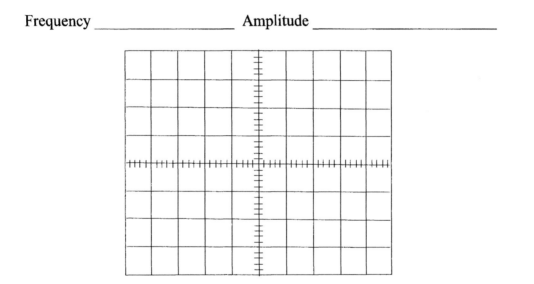

Plot 16-2 Three-stage ring oscillator.

4. The sawtooth oscillation you see is actually made of truncated portions of curved waveforms like you would see if you applied a square wave to an *RC* filter. As more stages are added to the ring, the time between transitions is increased and you will see more of a complete transition in each segment.

    Change the number of inverting gain stages to 7 with the same X10 gain as before. Figure 16-10 shows the circuit. Download the file to the PAM. You should see another sawtooth waveform on your oscilloscope. Notice the frequency and amplitude. Sketch the waveform in Plot 16-3.

Frequency _____ Amplitude _____

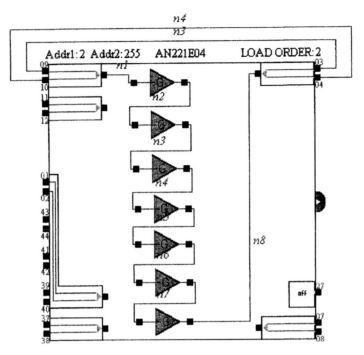

Figure 16-10 Seven-stage ring oscillator.

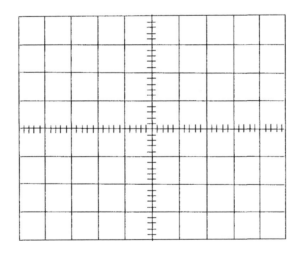

Plot 16-3 Seven-stage ring oscillator.

5. Although more stages slowed down the oscillator, you can slow it even more with an integrator and high-pass filter. Delete four of the gain stages, leaving the original three stages. Add an integrator (*Integrator*) CAM, accepting the default options but set the Integration Constant to 0.04. Add a biquadratic filter (*FilterBiquad*) CAM and select the high-pass and automatic filter topology options. Under Parameters, set the Corner Frequency to 8 kHz, the Gain to 1.5 and leave the Quality Factor at the default of 0.707 (this forms a Butterworth response). Add a gain limiter stage, setting the Gain to 20 and the Output Gain Limit to 1 V. Wire all of the CAMs in series. Your configuration should look like Figure 16-11.

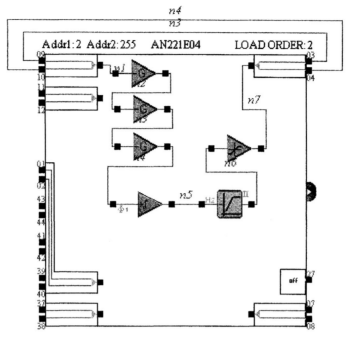

Figure 16-11 Special ring oscillator.

6. Download the file to the PAM. In Plot 16-4, measure the frequency and amplitude. Sketch the output waveform. Label the axes.

Frequency _____ Amplitude _____

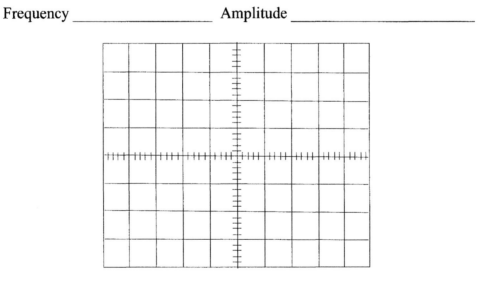

Plot 16-4 Waveform from the special ring oscillator.

7. Change the integration time constant on the integrator.
Observations:

_____

_____

8. On the right outside of the AD2 design window is a half-circle icon that when you click on it shows the resource meter. The resource meter shows how much of the internal resources of ASP are available and how much power is being dissipated. Open the resource meter and notice the little symbols for the capacitors, amplifiers and integrators arranged in four groups. When an ASP resource has been used, the symbol representing it is shown in a bright color. At the top of the resource meter there is also listed the availability of the Lookup Table (LUT) and the Counter resource that will be used together for the arbitrary waveform generator (AWG) that you will study in Part 3.
Observations:

_____

_____

Conclusion: Part 1

Questions: Part 1
1. To calculate the delay time for a signal through one inverting gain stage in a ring oscillator, you need to divide the period of the oscillations by two times the number of stages. Explain why.

2. Based on the measured frequency of oscillations in step 4, determine the delay time for one inverting gain stage.

3. How would you make a ring oscillator using CAMs with exactly four gain stages?

4. What information is available in the resource meter? (Hint: Try the Help > Help Topics > Index in AD2.)

Part 2: The Sine Wave Oscillator

As stated in the text box at the beginning of this experiment, the Anadigm ASP is a clock-driven system. When the analog signal frequency is low compared to the clock frequency (as in most previous experiments), you may not have observed the numerous steps that make up the signal. In this part, you will increase the analog signal frequency to be a major fraction of the clock frequency so that the individual steps are more visible. In extreme cases, the waveform is barely recognizable. The Anadigm ASP is still functional at these extremes but the signal-processing accuracy is reduced.

1. Start a new instance of AD2. Place a sine wave oscillator (*OscillatorSine*) CAM in the design window. In the *Set CAM Parameters* window, set the Osc. Frequency to 800 kHz and the Peak Amplitude to 2.0 V. Note that the ClockA frequency is at the default value of 4000 kHz.

    Notice that a Warning message appears in the parameter-setting window as shown in Figure 16-12. This message indicates a functional limitation of the *OscillatorSine* CAM at its 800 kHz maximum frequency setting. The potential actions you can take are to accept the limitation, to try a higher frequency clock setting, or to use a lower oscillator frequency where this warning does not apply. Try selecting the ClockA for 16000 kHz. The result is that you see a different Warning message: "It is not recommended to run this CAM's clock at a frequency greater than 4 MHz." This is a more serious warning so put ClockA back to the 4000 kHz setting and try a lower signal frequency. Set the oscillator frequency to 400 kHz where no warning messages occur.

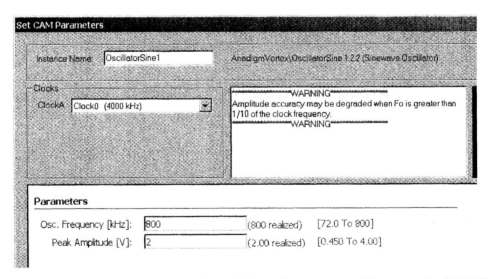

Figure 16-12 Warning message issued for a sine wave oscillator set for 800 kHz.

2. Add an inverting gain (*GainInv*) CAM to the design window and accept the default gain of 1 to act as a buffer. Connect the output to OutputCell1 and put a simulated scope probe on the output of the inverting stage as shown in Figure 16-13.

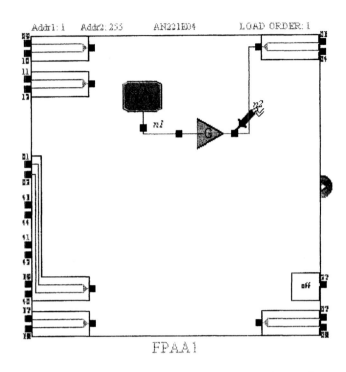

Addr1: 1　　Addr2: 255　　　AN221E04　　　　LOAD ORDER: 1

FPAA1

Figure 16-13 Sine wave oscillator with inverting gain stage.

3.　In the *Simulate* menu, choose Setup Simulation… and set the End Time to 100 μs.
Run the simulation. On the simulated scope display, set the Time Per Div to
500 ns and push the time slider full to the right to show two valleys and two peaks
of the simulated signal as in Figure 16-14.

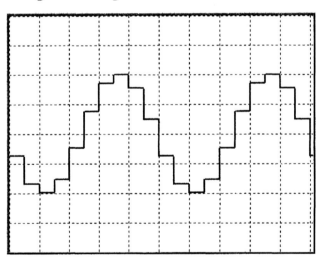

Figure 16-14 Sine oscillator CAM output at 400 kHz setting and 4000 kHz clock

4.　The signal-processing clock is 4000 kHz so this 400 kHz sine wave is processed
as 10 steps per cycle. This waveform represents the sine wave, which looks much
better after applying a smoothing filter. Download the configuration into your
PAM and view the signal at OUT1. The output buffer shown in Figure 14-11
includes a 400 kHz low-pass "smoothing filter," which will also reduce the output
a little. Sketch the signal in Plot 16-5. Label the axes.

230

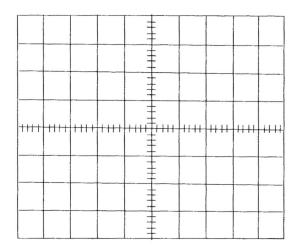

Plot 16-5 Oscillator signal at OUT1 of PAM.

5.   Place a second sine wave oscillator in the lower part of the design window and set
the frequency to 7 kHz. After entering "7" and clicking outside the *Osc. Frequency*
box, you will see that it has changed to a red background color indicating an out-of-
range error condition as shown in Figure 16-15. Notice that the "40 kHz realized"
value is not the value you want.

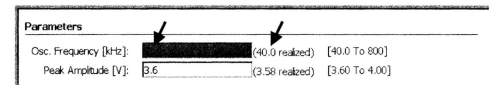

Figure 16-15 Out of range indication for *OscillatorSine* parameter setting

For the *OscillatorSine* CAM, there is a limited range of frequencies that
can be generated for a given clock frequency. The solution to this error is to try a
slower clock. Try changing ClockA to 2000 kHz. You should still see the out-of-
range error. Now try the slowest available clock frequency, which is 250 kHz as
shown in Figure 16-16. This results in no error indicators or warning messages.

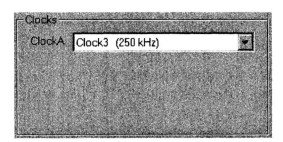

Figure 16-16

231

6. Change the Peak Amplitude in the Parameters setting from 3.6 V to 2.0 V. Close the parameter-setting window.

7. Add an inverting gain stage with the gain of 1 in series with the oscillator and connect the output to OutputCell2 as shown in Figure 16-17. Notice that the wire connecting the new *OscillatorSine* CAM to the *GainInv* CAM is dashed. Point your cursor to this dashed line and hold it there a moment while an error report window opens up that says:

    "Design rule issues: There is a sample clock mismatch on this wire"

    It is a very good practice to have all CAMs that are processing the same signal operating at the same clock frequency. The solution is to set the downstream CAM (the second *GainInv* CAM) to the 250 kHz setting. Double-click the inverting amplifier and change the clock frequency to 250 kHz. When you have done this, notice that the wire is now drawn as a solid line.

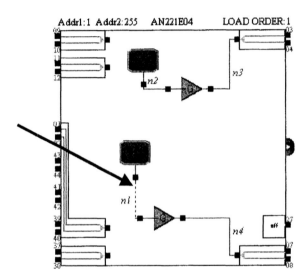

Figure 16-17 Dashed line indicating clock frequency mismatch between CAMs.

8. Move the simulator scope probe to the input of the lower *GainInv* CAM. Set the Simulation End Time to 800 μs and run the simulation. Set the Time Per Div to 20 μs and move the time slider to about the 70% point to show two sine wave peaks on the display. You can clearly see the sample steps in the sine wave.

    Look very carefully at the rising edge of each sine wave at the zero crossing and notice that the amplitude of the first step after the zero crossing to the left is not the same as the first step to the right (adjust the time slider as needed to see this). The reason is that there are 250,000 clock cycles per second processing a 7000 cycles per second signal, which implies 35.71 steps per cycle of the sine wave. The simulation is static (unchanging), but this is not what you will see on the PAM with your oscilloscope because it is constantly changing.

    Download this configuration into your PAM. Move your oscilloscope probe to OUT2 and set the scope for 1 V/Div and about 20 μs/Div. When the configuration is loaded into the PAM, you can see the uneven number of steps per cycle on the oscilloscope as it continuously redraws the waveform. (Put the scope in Sample mode to see this, not Average mode, if your scope has this feature.) Draw the waveform on Plot 16-6. Label the axes.

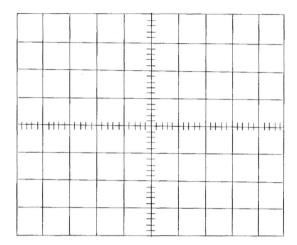

Plot 16-6 A 7 kHz sine wave using the internal 250 kHz clock.

9. Change the lower *OscillatorSine* CAM to 41 kHz, which will give 6.0976 clock steps for each cycle of the sine wave. Note that the Amplitude accuracy warning message appears. Set the Simulation End Time to 400 μs and run the simulation. Set the Time Per Div to 5 μs and move the time slider to the right to show two sine wave peaks on the simulation display. Note the amplitude differences in the first positive step after the zero crossing for each displayed cycle.

10. Set your oscilloscope to 5 μs/Div and then download this configuration into the PAM. Draw the waveform in Plot 16-7 that you see on OUT2. Label the axes.

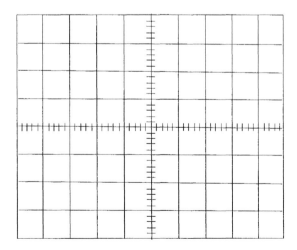

Plot 16-7 A 41 kHz sine wave using the internal 250 kHz clock.

Pushing the Anadigm ASP to its limits produces interesting waveforms as you have seen. These waveforms can be useful if you can apply a smoothing filter appropriate for the signal frequency and where the amplitude errors are not important.

Conclusion: Part 2

Questions: Part 2

1. As stated in the introduction, the Anadigm ASP operates with samples of the analog signal.
    (a) Why do you think that 5 samples per cycle is the lower limit chosen to be implemented in the AD2 program for all of the ASP CAMs?

    (b) What is the absolute lowest number of samples per cycle that can be taken and still know at least the frequency of an analog signal? (*Hint*: Look up Nyquist frequency on the Internet.)

2. Explain what is done to the ASP signal by the downstream *smoothing filter* in the PAM.

3. What do you think could happen to the signal at OUT2 if you had ignored the clock frequency mismatch indication (the dashed line) in Step 7 and downloaded the configuration into the PAM?

Part 3: The Arbitrary Waveform Generator

As explained in the introduction, an arbitrary waveform generator (AWG) is a standard laboratory instrument that can produce a periodic wave that is previously defined in a lookup table. For AD2, the AWG is named the *PeriodicWave* CAM. It works from a table of values, which are presented in sequential order to a digital-to-analog converter (DAC) repeatedly. The Anadigm ASP uses ClockA to acquire values for the DAC from the Lookup Table and ClockB to drive the DAC as in Figure 16-18.

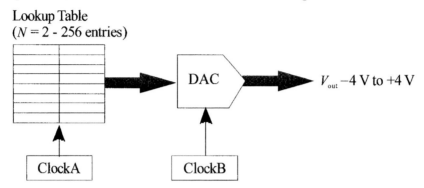

Figure 16-18 Block diagram of arbitrary periodic waveform generator.

1.   Start a new AD2 instance. Place an arbitrary waveform generator (*PeriodicWave*) CAM in the design window. In the *Set CAM Parameters* window shown in Figure 16-19, notice that there are two different clocks, a Lookup Table button, and the *Counter Reset Value* (which will be set to one less than the number of table entries). ClockA drives the Lookup Table and must be slower than ClockB by an integer multiple that we will call K.

$$\text{ClockB} = K * \text{ClockA}$$

K is constrained within the design of the AWG CAM to be 5 or greater. You will use 8 to begin with to create a simple repetitive waveform.

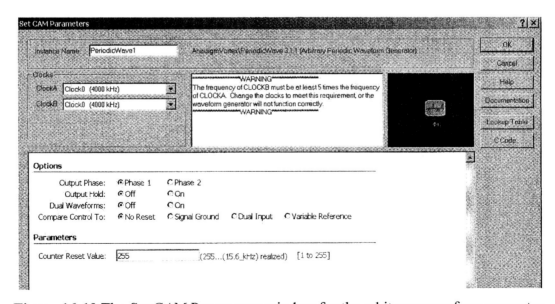

Figure 16-19 The *Set CAM Parameters* window for the arbitrary waveform generator.

2. Select ClockB first (the faster analog signal-processing clock) to be 16000 kHz and ClockA (the slower Lookup Table Clock) to be 2000 kHz for a K value of 8. Notice that the Warning message about clock frequencies goes away.

3. Now select the Lookup Table button to bring up the *Lookup Table – Output Voltage Sequence* window as shown in Figure 16-20. You can enter values into the Lookup Table (LUT) in this window in either of two ways:
    *   Highlight each line and type in a new output voltage (−4.000 to +4.000) in place of the 0.000 entry.
    *   Load a previously saved .CSV (comma delimited file) of sequential values from a spreadsheet with the Load button.

    Using the first method, enter 1.000 into the first 6 lines, leave the next 6 lines as zeroes and then enter −1.000 into the next 6 lines. Taking the next 6 lines as all zeroes, you will have an $N = 24$ entry sequence of +1 V, 0 V, −1 V, 0 V at a repetitive frequency of ClockA / 24 = 83.33 kHz. Click Apply and then OK. Close the LUT value entry screen and you will return to the *Set CAM Parameters* window.

Index	Requested (-4 to 4)	Realized
0	1.000	1.000
1	1.000	1.000
2	1.000	0.000
3	1.000	0.000
4	1.000	0.000
5	1.000	0.000
6	0.000	0.000
7	0.000	0.000
8	0.000	0.000
9	0.000	0.000
10	0.000	0.000
11	0.000	0.000
12	-1.000	0.000
13	-1.000	0.000
14	-1.000	0.000
15	-1.000	0.000
16	-1.000	0.000
17	-1.000	0.000
18	0.000	0.000
19	0.000	0.000
20	0.000	0.000
21	0.000	0.000
22	0.000	0.000
23	0.000	0.000

OK Apply Cancel Load Save

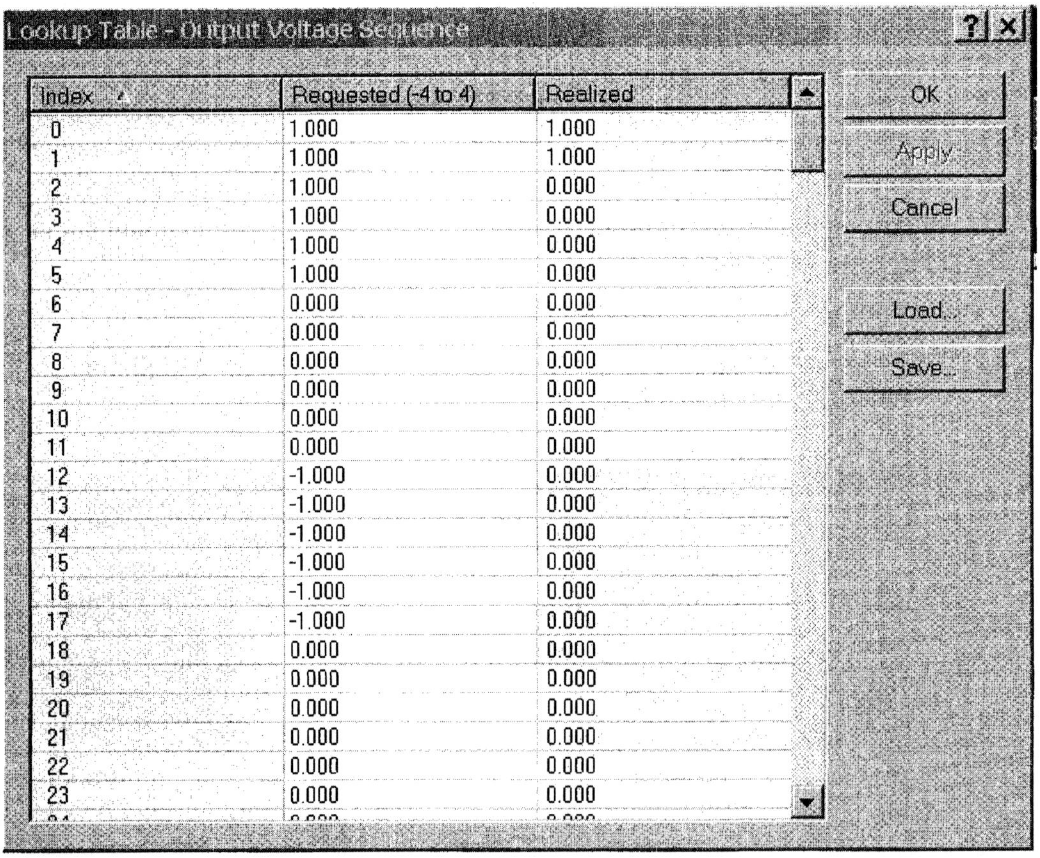

Figure 16-20 AWG Lookup Table value entry screen.

4. In the *Set CAM Parameters* window, enter 23 for the *Counter Reset Value*. The *Counter Reset Value* should always be set to one less than the number of LUT entries *N*. Select the *Output Hold* option as ON and then close the window.

5. Wire the output of the AWG to OUT1. Install a simulation scope probe to the output of the AWG per Figure 16-21. In the *Simulate* menu, choose Setup Simulation… and set the End Time for 100 μs. Run the simulation. On the simulated oscilloscope, set the Time Per Div to 5 μs and observe this alternating high / zero / low / zero waveform commonly seen in telephone circuitry. Measure the frequency and record the waveform in Plot 16-8. Label the axes.

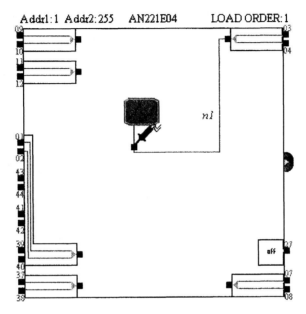

Figure 16-21 Arbitrary Periodic Waveform CAM

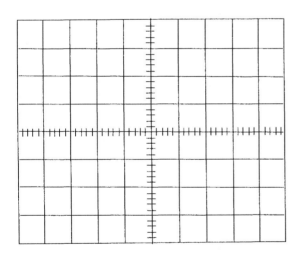

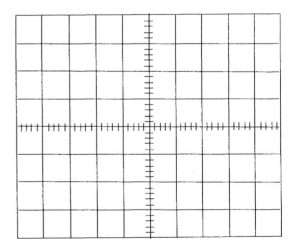

Plot 16-8 Simulated waveform.

Frequency: Simulated _____

Plot 16-9 Measured waveform.

Measured _____

6. Download the AWG configuration into your PAM. Measure the frequency and record the waveform in Plot 16-9. Why do you think the simulated and measured waveforms are different? (*Hint*: The PAM output is not taken directly from the ASP.)

_____

_____

7. Select the Lookup Table Clock (ClockA) to be 250 kHz, which should give a waveform repetitive frequency of 250 kHz / 24 = 10.42 kHz. Download to the PAM, measure the frequency, and observe the waveform.

   Frequency: Actual _____

   Other than the frequency, what is clearly different between this waveform and the waveform observed in Plot 16-9?

_____

_____

8. In this step you will create a triangle wave with the AWG using values from a spread sheet program that can be loaded directly into the LUT. Open a spreadsheet (like Microsoft Excel) and enter the value −3.8 in the A1 cell (all values will go into column A). Enter −3.7 in the A2 cell. Highlight these two cells and then point your cursor exactly on the little black cross in the lower right "picture frame" surrounding these highlighted cells. Click and drag down column A as Excel fills in the values with the +0.1 increment implied by these two cell values. Stop at A77 where the value is +3.8 V. Enter 3.7 into A78, highlight A77 and A78 (indicating −0.1 increment) and then click and drag further down the column until −3.7 is reached at A152 ($N = 152$ for this waveform). Now save this file selecting the "CSV (Comma delimited) (*.CSV)" file type and accept all the qualifiers that Excel offers in saving the file. Find the two entries which are close approximations to zero (cell A39 and A115), enter exactly zero and then save the file again. (This fixes a quirk in Excel.)

9. Return to the AD2 file you were using and open the *Set CAM Parameters* window for the AWG CAM. Click the Lookup Table button, which will put you back to the *Lookup Table – Output Voltage Sequence* window. You can now enter values into the LUT. Click the Load button. Locate the .CSV file that you just saved and click the Open button to load the data from this file into the Anadigm AWG CAM. Click Apply and then OK to close the window to return to the *Set CAM Parameters* window.

10. In the *Set CAM Parameters* window, set the *Counter Reset Value* to 151 (1 less than *N*) and *Output Hold* = ON. Select ClockA (the Lookup Table Clock) to 250 kHz for a K value of 64. With a simulator scope probe at the output of the AWG icon, set the simulator *End Time* for 1 ms and then run the simulation. Set the Time Per Div on the simulated scope display to 100 μs. You should see a little more than 1½ cycles of the triangle wave that you have just defined as illustrated in Figure 16-22.

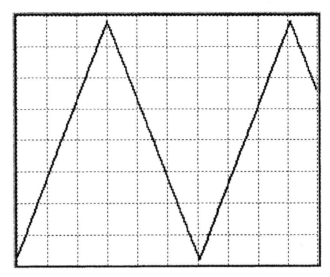

Figure 16-22 Triangle waveform from AWG CAM.

11. Measure the frequency and period of the triangle waveform on the simulation.
Frequency = _____   Period = _____

12. Download this configuration into your PAM and observe the waveform at OUT1. Compare the observed waveform with the simulated waveform. Is the frequency and period the same as the simulation?
Observations:
_____

_____

13. Double-click the AWG to open the *Set CAM Parameters* window. Increase the frequency of ClockA to 2000 kHz. Download this configuration into your PAM and observe the waveform at OUT1.
Observations:
_____

_____

14. Another useful application of the AWG CAM is to create a precise DC voltage for a reference. To create a DC reference voltage, open the *Set CAM Parameters* window for the AWG CAM and click the Lookup Table button. In the *Lookup Table – Output Voltage Sequence* window, set all the previous entries to zero value (or instead you could start with a new instance of the AWG CAM). Set the first two entries in the LUT to the desired value (use –3.46 for this example). Click Apply and then OK to return to the *Set CAM Parameters* window. Enter 1 for the *Counter Reset Value* (you cannot enter zero, which is why you need at least two entries in the LUT). Select *Output Hold* = ON and close the window. Run the simulation. Turn on the Cursor feature of the simulated scope and note the voltage for Channel 1.

Voltage from the simulation: _____

Download this configuration file into your PAM and measure the voltage at OUT1.

Voltage at OUT1: _____

Why isn't the measured value exactly the same as the simulated value?

_____

Conclusion: Part 3

Questions: Part 3
1. In Step 11, you measured the frequency of the triangle waveform. Prove that the frequency you measured is as expected based on the number of entries in the LUT and the clock frequency. (*Hint*: See Step 3 for more information.)

2. In your everyday activities, what common item do you use that is conceptually similar to the arbitrary waveform generator, i.e. a stream of digital values presented in sequential order to a DAC?

3. In Step 13, you changed ClockA to 2000 kHz. Why did this action change the output frequency?

Experiment 17 Voltage Regulators

Regulated power supplies are designed to provide a constant dc voltage to a changing load from a source of alternating current (ac). Good regulation requires that the output voltage is constant for variation in line voltages, load resistance, or temperature change. Power supplies require heat sinks to get rid of unwanted heat. For this experiment, the circuits are operated with low power to avoid the need for heat sinks but still illustrate the ideas.

    The basic series regulator circuit places the pass transistor in series with the output; hence the term *series regulator*. This circuit is one of the most popular forms of regulators. In Part 1, you will construct and test a series regulator and connect it to a bridge rectifier circuit, forming a complete regulated power supply.

    In Part 2, you will test a three-terminal regulator. The circuit is similar to the Application Activity but uses the 7805 regulator. The 7805 is used to set a fixed +5 V in small applications; however, it can also serve as a simple current source. The supplied current is nearly constant despite changes in the input voltage or load as you will see.

Reading

Floyd, *Electronic Devices*, Eighth Edition, Chapter 17

Key Objectives

    Part 1: Construct and test a voltage regulator circuit. From measured voltages, compute the line and load regulation for the circuit.

    Part 2: Construct and test a power supply and a current source featuring a 7805 IC.

Components Needed

Part 1: The Series Regulator

Resistors: four 330 Ω resistors, one 1.0 kΩ, one 1.2 kΩ, one 2.7 kΩ
One 1 kΩ potentiometer
Transistors: one 2N3904 *npn* transistor, one 2N3053 *npn* power transistor
Diodes: one 1N4733A 5 V zener diode, four 1N4001 rectifier diodes
One 1000 µF capacitor

Part 2: IC Regulators

Resistors: one 330 Ω, one 2.2 kΩ
One 12.6 V ac transformer with fused line cord
Four diodes 1N4001 (or equivalent)
Capacitors: one 220 µF, one 0.01 µF capacitor
One 7805 or 78L05 regulator
Two red LEDs
One ammeter capable of measuring 20 mA

Part 1: The Series Regulator

Safety Note: The power dissipated in the pass transistor for this experiment will cause it to become hot, but a heat sink is not required for the loads specified.

1. Measure and record the values of the resistors listed in Table 17-1.

2. Construct the series regulator circuit shown in Figure 17-1. This circuit is limited to relatively small power levels for illustrating regulator operation unless heat sinking of the pass transistor is provided. Connect the input (V_{IN}) to a dc power supply set to +18.0 V.

Table 17-1

Resistor	Listed Value	Measured Value
R_1	2.7 kΩ	
R_2	330 Ω	
R_3	1.0 kΩ	
R_4*	1 kΩ	
R_5	1.2 kΩ	
R_L	330 Ω	

*potentiometer; record maximum resistance

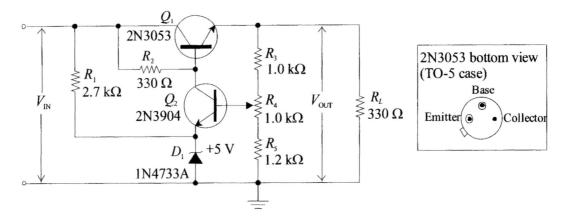

Figure 17-1

3. Compute the minimum and maximum output voltage for the regulator. The minimum voltage is found by assuming the $V_{BASE(Q2)}$ is 5.7 V (the zener drop + 0.7 V). This voltage is dropped across R_4 and R_5. Because R_3, R_4, and R_5 are in series, you can find the output by the proportion:

$$\frac{V_{OUT(min)}}{V_{BASE(Q2)}} = \frac{R_3 + R_4 + R_5}{R_4 + R_5}$$

Enter the computed $V_{OUT(min)}$ in Table 17-2. Set up a similar proportion to find $V_{OUT(max)}$. Enter the computed values in Table 17-2.

Table 17-2

Parameter	Computed Value	Measured Value
$V_{OUT(min)}$		
$V_{OUT(max)}$		

4. Maintain the input voltage at +18.0 V. Test the minimum and maximum output voltage by varying R_4 over its range. Enter the measured values in Table 17-2.

5. In this step and step 6, you will measure the variation on the output voltage due to a change in the input voltage and compute the line regulation. With the input voltage at +18 V, adjust R_4 for an output of +10.0 V. (This is shown as the starting value in Table 17-3.) Then set the input voltage to each value listed in Table 17-3, measure and record the output voltage. (Note that a ± 4 V input variation on the input is more than would be observed in normal operation).

Table 17-3

V_{IN}	V_{OUT} (measured)
+18.0 V	+10.0 V
+17.0 V	
+16.0 V	
+15.0 V	
+14.0 V	

Table 17-4

Step	Quantity	Measured Value
6	Line regulation	
7	V_{NL}	+10.0 V
	V_{FL}	
	Load regulation	
8	$V_{ripple(in)}$	
	$V_{ripple(out)}$	

6. Compute the line regulation from the data taken in step 5. The line regulation is given by the following equation:

$$\text{Line Regulation} = \frac{\Delta V_{OUT}/V_{OUT}}{\Delta V_{IN}} \times 100\%$$

Use the first and last entry of Table 17-3 to compute the line regulation and enter the value in the first row of Table 17-4.

7. Find the load regulation by determining the change in the output voltage between no load and full-load and dividing by the output voltage at full load. Load regulation is usually expressed as a percentage. For this circuit, assume the full-load output current is 90 mA (This is lower than most supplies but helps to keep heat down). Remove the 330 Ω load resistor and set the input (no-load) voltage to +16 V. Adjust the output voltage to +10.0 V (V_{NL}). Then install three parallel 330 Ω resistors across the output and measure the output load voltage (V_{FL}). Compute the load regulation from the equation:

$$\text{Load Regulation} = \frac{V_{NL} - V_{FL}}{V_{FL}} \times 100\%$$

Enter the measured full-load voltage, and the computed load regulation in Table 17-4.

8. Figure 17-2 shows a complete regulated power supply that has bridge rectifier input. (The bridge rectifier is the same circuit that you studied in Part 1 of

243

Experiment 2 and you will see how the regulator can reduce the ripple voltage.) Disconnect the power supply that you have been using as an input device and add the bridge rectifier circuit shown in Figure 17-2. The load consists of the three 330 Ω resistors in parallel from step 7. Measure the peak-to-peak ripple voltage across C_1 ($V_{ripple(in)}$) and across the output load ($V_{ripple(out)}$). Couple your oscilloscope with ac coupling to view the ripple. Record the results in Table 17-4.

Multisim

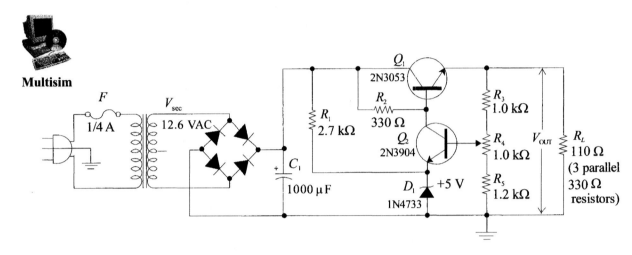

Figure 17-2

Conclusion: Part 1

Questions: Part 1

1. What change would you suggest to the circuit in Figure 17-2 if you needed to reduce the ripple?

2. The major drawback to a series regulated supply is inefficiency. The efficiency is defined as the percentage of input power that can be delivered to the load. For the circuit in Figure 17-2, assume the input power is 2.01 W (17.5 V at 115 mA). With the output set to 10 V and a 110 Ω load resistance, compute the efficiency of the regulator.

Part 2: IC Regulators

1. The pinouts for the 7805 and the 78L05 regulator are shown in Figure 17-3; one of these will be used in this experiment.

    A bridge rectifier circuit with an IC regulator is shown in Figure 17-4. The circuit is similar to the one in Experiment 2, Part 1, but with the addition of the three-terminal IC regulator. The transformer has a 12.6 V_{rms} secondary. Notice that <u>no</u> terminal on the transformer is grounded because of the bridge configuration. The center-tap (if present) is not connected.

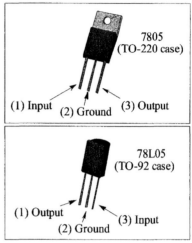

Figure 17-3

The input to the regulator is first filtered by the 220 μF filter capacitor. The 0.01 μF capacitor provides a path for noise to ground to reduce any "spikes" in the output. Connect the circuit, and measure the output ripple. If you did Experiment 2, Part 1, compare your result with the unregulated bridge circuit you tested in step 9 of that experiment (note differences in capacitors).

Observations:

_____

_____

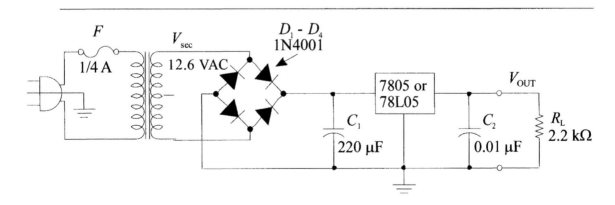

Figure 17-4

2. A three-terminal regulator used as a current source is shown in Figure 17-5. Use either the 7805 or 78L05 regulator. Connect the input to a variable dc supply, so that you will be able to change the input. Vary the input voltage between +10 V and +16 V. Measure the output current for each input voltage listed in Table 17-5 using a meter that can read a minimum of 20 mA full scale. Enter the measured current in the table.

Observations:

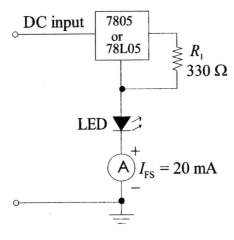

Figure 17-5

Table 17-5 Data for one LED

Voltage	Current
10 V	
12 V	
14 V	
16 V	

Table 17-6 Data for two LEDs

Voltage	Current
10 V	
12 V	
14 V	
16 V	

3. Add a second LED of the same color in parallel with the first LED. Repeat the measurements from step 2 with this load. Enter the data in Table 17-6.

Observations:

Conclusion: Part 2

Questions: Part 2

1.   The current source in Figure 17-5 has more current than you would predict by dividing the output voltage by the resistance of R_1. Why?

2.   How would you change the circuit in Figure 17-5 to produce a current of 7.5 mA?

Multisim Simulation

Multisim

Multisim files for the lab manual are on the website www.prenhall.com/floyd. Open the Multisim file Experiment_17_Regulators-nf. The circuit is a simulation of the one in Figure 17-2. A 2 Ω surge resistor is added to prevent the fuse from blowing on initial turn on. Check the range of voltage on the output (using the multimeter) and measure the ripple voltage with the oscilloscope. Determine the maximum power dissipated in the series pass transistor. Compare the simulation with the experimental results.

After testing the regulator, open the file Experiment_17_Regulators-fault and troubleshoot the circuit.

Experiment 18 Communications Circuits

Tuned amplifiers are widely used in high-frequency communication systems and in instruments such as the spectrum analyzer. Several intermediate frequency (IF) amplifiers usually account for most of the gain in the system. As explained in the text, a radio frequency (RF) is converted to a lower frequency by mixing the RF with a local oscillator. The oscillator frequency is variable, and "tracks" the incoming signal at a fixed difference frequency producing the IF which is fixed.

In Part 1, you will construct and test a transformer-coupled IF amplifier that is designed to amplify frequencies around 455 kHz, a standard IF frequency. The transformer in the IF amplifier has a small capacitor (about 180 pF) in parallel with the primary, forming a parallel resonant circuit. The entire assembly is mounted inside a shielded metal enclosure that in normal operation is grounded. The circuit is tuned by using a nonmetallic screwdriver to adjust a tuning slug in the core.

In Part 2, you will investigate an integrated circuit phase-locked loop (PLL) circuit, the 565. The 565 PLL is composed of a phase comparator, a low-pass filter and amplifier, and a voltage-controlled oscillator (VCO) as shown in Figure 18-1.

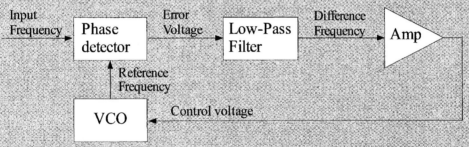

Figure 18-1

The input frequency is compared with the VCO frequency and the sum and difference frequencies are generated by the phase comparator. The higher (sum) frequency is filtered out and the difference frequency is converted to a control voltage for the VCO. This control voltage causes the oscillator to change in the direction of the input frequency. As a result, the VCO can "track" any change in the input frequency but produces a "clean" signal that can track a noisy input or even generate a different shaped waveform from the input.

Reading

Floyd, *Electronic Devices*, Eighth Edition, Chapter 18

Key Objectives

Part 1: Construct an intermediate frequency amplifier and measure its dc and ac parameters including gain, bandwidth, and *Q*.

Part 2: Test a phase-locked loop (PLL) to determine the free-running frequency, the capture and lock range, and the voltage as a function of frequency on the FM output pin. Use the PLL as a frequency multiplier.

Components Needed

Part 1: The IF Amplifier

Resistors: one 220 Ω, one 470 Ω, one 4.7 kΩ, two 10 kΩ, one 56 kΩ
Capacitors: two 0.1 μF
Transistors: two 2N3904 npn (or equivalent)
3rd stage IF transformer (20 kΩ primary to 5 kΩ secondary) Mouser 42IF303
Optional: Frequency counter (An oscilloscope can substitute.)

Part 2: The Phase-Locked Loop

Resistors: three 1.0 kΩ, one 2.0 kΩ, one 10 kΩ
Capacitors: one 1000 pF, one 2200 pF, one 0.1 μF, one 1.0 μF
One 10 kΩ potentiometer
One 2N3904 *npn* transistor (or equivalent)
One LM565 phase-locked loop
One 7490A decade counter
Optional: frequency counter (An oscilloscope can substitute.)

Part 1: The IF Amplifier

1. Measure and record the values of the resistors listed in Table 18-1.

Table 18-1

Resistor	Listed Value	Measured Value
R_1	56 kΩ	
R_2	4.7 kΩ	
R_3	10 kΩ	
R_{E1}	220 Ω	
R_{E2}	470 Ω	
R_L	10 kΩ	

Table 18-2

DC Parameter	Computed Value	Measured Value
V_B		
V_E		
I_E		
V_C		
V_{CE}		

2. Compute the dc parameters listed in Table 18-2 for the IF amplifier shown in Figure 18-2. V_B, V_E, and V_C are referenced to ground. Because of the low dc resistance of the transformer primary (about 5 Ω), assume that $V_C = V_{CC}$. Enter your computed values in Table 18-2.

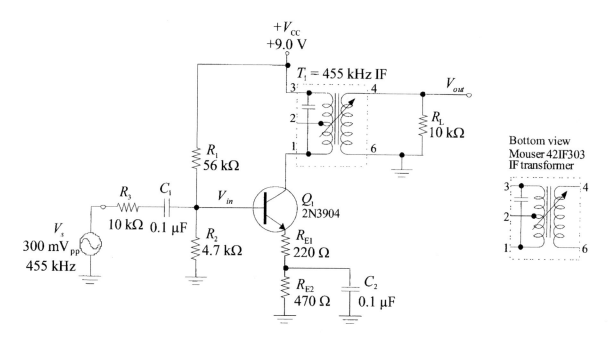

Figure 18-2

3. Construct the amplifier shown in Figure 18-2. The signal generator should be turned off. Measure and record the dc voltages listed in Table 18-2.

4. The input signal, V_{in}, is measured at the transistor's base. This input voltage will *increase* at resonance due to the reduced loading on the generator and will depend to some extent on the exact frequency of the resonant circuit. Set the function generator for a 300 mV$_{pp}$ sinusoidal wave at 455 kHz. A low capacitance probe should be used for measurements to avoid probe loading effects. While observing the generator, adjust the frequency (slightly) for maximum amplitude (or tune the transformer for maximum amplitude). You will need to readjust the level for the 300 mV$_{pp}$ value at the point where the frequency peaks. This is both V_{in} and the ac base voltage, V_b. Record the measured value as V_b in Table 18-3.

5. The collector voltage and the gain depend on the exact Q of the primary resonant circuit, the amount of coupling between the primary and secondary coils, the load on the secondary, and the ac resistance of the emitter circuit. In addition, loading effects from the measuring instrument can affect the results and this is why a low-capacitance probe was suggested. Measuring the primary side of the transformer will produce a greater loading effect than the secondary; however, it is necessary to view the primary to measure the gain. Measure the remaining ac parameters listed in Table 18-3. The gain, A_v, is the ratio of the ac collector voltage to the ac

base voltage, V_b. The output voltage is measured at the secondary of the transformer (across the load resistor).

Table 18-3

AC Parameter	Measured Value
V_b	
V_c	
A_v	
$V_{out(tot)}$	

Table 18-4

AC Parameter	Measured Value
f_c	
f_{cu}	
f_{cl}	
BW	
Q	

6.   In this step, you will measure the center frequency of the IF amplifier. It is much easier to measure it accurately if you have access to a frequency counter. Like the scope, the frequency counter can load the circuit; it will have the smallest effect if used on the secondary side.
      While observing the output voltage on an oscilloscope, adjust the frequency for the maximum output. Then adjust the oscilloscope so that the peak-to-peak output voltage covers exactly 5 vertical divisions. You will probably have to take the oscilloscope out of vertical calibration to set this level exactly. Measure and record the frequency of the maximum output using a frequency counter (or the oscilloscope if a frequency counter is unavailable). Record this as the center frequency, f_c, in Table 18-4.

7.   Measure the bandwidth (BW) of the IF amplifier. Again, a frequency counter is the best way to measure the frequency accurately. Raise the generator frequency slowly while observing the output on an oscilloscope. Adjust the frequency until the peak-to-peak output voltage indicates 3.5 divisions (70%). The frequency above the center frequency at which the output is 70% of the maximum is the upper cutoff frequency (f_{cu}). Record this upper cutoff frequency in Table 18-4.

8.   Adjust the generator frequency for the lower cutoff frequency by watching for the 70% point below the center frequency. Record this as the lower cutoff frequency, f_{cl}, in Table 18-4.

9.   Compute the BW of the circuit by subtracting the lower cutoff frequency from the upper cutoff frequency. Compute the Q of the circuit by dividing the center frequency by the BW. Enter the BW and Q in Table 18-4.

Conclusion: Part 1

Questions: Part 1

1. If you wanted to determine if the oscilloscope probe had a loading effect on the circuit, you could connect a second identical probe to the same point in the circuit. Explain how this would allow you to see if probe loading was a factor.

2. Explain why the voltage gain of the CE amplifier in this experiment was highest at the resonant frequency.

3. Assume a student determined that the voltage from the generator was 400 mV$_{PP}$ and the drop across R_3 was 100 mV$_{PP}$. For this amplifier, what is the input resistance?

4. If you measured 0 Vdc on the collector, which of the following could account for the problem?
   (a) open primary
   (b) open secondary
   (c) open R_1
   (d) open R_{E1}
   (e) power supply off

Part 2: The Phase-Locked Loop

1. Measure and record the values of the components listed in Table 18-5. If you cannot measure the capacitors, enter the listed values.

Table 18-5

Component	Listed Value	Measured Value
R_1*	2.0 kΩ	
R_2	1.0 kΩ	
R_3	1.0 kΩ	
C_1	2200 pF	
C_2	0.1 μF	
C_3	1000 pF	

* plus 10 kΩ potentiometer in circuit

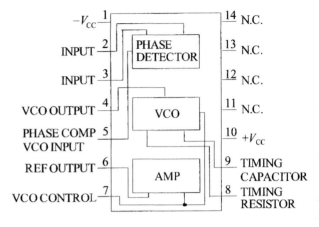

Figure 18-3

2. For reference, the LM565 pinout is shown in Figure 18-3. The circuit in Figure 18-4 shows the LM565 phase-locked loop in a test circuit. Notice that the VCO output is returned to the phase detector by the jumper from pin 4 to 5. The VCO has a free-running frequency determined from the equation:

$$f_0 = \frac{0.3}{R_1 C_1}$$

Compute the free-running frequency range by assuming R_1 is set to the minimum and to the maximum value. Enter the computed free-running frequencies in Table 18-6.

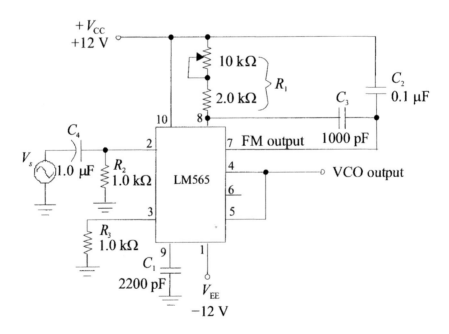

Figure 18-4

3. Construct the circuit, but leave the signal generator turned off or disconnected. Connect a scope (or frequency counter) to pin 4 and measure the free-running frequency, f_0, of the VCO with R_1 set to the minimum and maximum values. Enter the measured frequencies in Table 18-6.

Table 18-6

Step	Quantity	Computed Value	Measured Value
2 and 3	free-running frequency, $f_{0(min)}$		
	free-running frequency, $f_{0(max)}$		
4	lower capture frequency	Range =	
5	upper capture frequency	± 1.9 kHz	
6	lower lock frequency	Range =	
	upper lock frequency	± 8.3 kHz	
	frequency in (multiplier)	2.5 kHz	
8	frequency out (multiplier)		

Capture Range

4. With the generator off, adjust R_1 for a free-running frequency of 25 kHz. Then turn on the generator and view it on Ch-1 of your oscilloscope. Trigger the scope from Ch-1. Adjust the generator for a 1.0 V$_{pp}$ signal at 20 kHz. Connect the Ch-2 probe to the VCO output (pin 4). You should see that the signal on Channel 2 appears to free run. *Slowly* increase the frequency of the generator while observing the two signals. As you approach 25 kHz, the VCO should lock onto the input frequency. This is the lower capture frequency. Record the measured value in Table 18-6.

   Note that the computed capture and lock ranges have already been entered into the table based on the following equations:

$$f_{lock} = \pm \frac{8 f_0}{V_{CC}} \qquad\qquad f_{cap} = \pm \frac{1}{2\pi} \sqrt{\frac{2\pi f_{lock}}{3.6 \text{ k}\Omega \times C_2}}$$

The free-running frequency will not be centered in these ranges.

5. Set the function generator for a frequency of approximately 30 kHz. This time *slowly* lower the frequency of the generator and watch for capture. Record the measured upper capture frequency in Table 18-6.

Lock Range

6. Note that once captured, you can reduce the frequency below the lower capture frequency and retain lock or raise the frequency above upper capture frequency and retain lock. Notice that the lock range is wider than the capture range. Capture the input signal and measure the lowest and highest frequency for which you can retain lock. Record the measured lock frequencies in Table 18-6.

7. When the signal is locked, observe the waveform on pin 7.

Observations:_____

Frequency Multiplier

8. Figure 18-5 shows the basic PLL circuit with a 7490A decade counter added to the feedback circuit. Add the 7490A and transistor to the circuit as shown. Notice that the 7490A operates only between +5.0 V and ground. The transistor prevents the input signal from exceeding this range. Set the input frequency for 2.5 kHz and observe the output on pin 4 of the LM565. The output should be locked to the input. The computed output frequency is the input frequency times the decade count ratio provided by the 7490A. Record this as the computed value of frequency out in Table 18-6. Then measure the input and output frequencies and record them in Table 18-6.

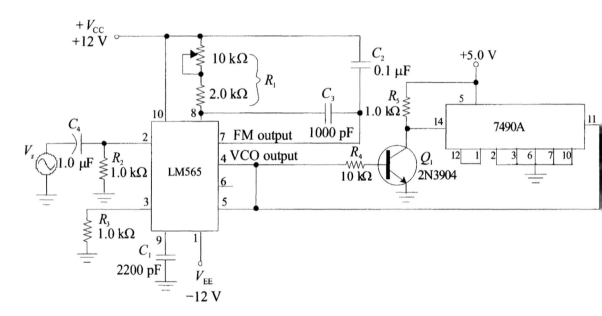

Figure 18-5

Conclusion: Part 2

Questions: Part 2

1. What change would you make to the circuit in Figure 18-4 to increase the free-running frequency by a factor of ten?

2. What is the purpose of the *npn* transistor in the multiplier circuit of Figure 18-5?

3. Explain the difference between the capture range and the lock range. Which is normally larger?

4. Assume you want to set up a phase-locked loop to multiply a frequency by 3.5. A divide by seven divider is used in the feedback loop. What must be done to the output frequency?

5. (a) Based on the manufacturer's specification sheet, what is the maximum power supply voltage for the 565?

    (b) What effect does this voltage have on the lock range?

Appendix A- List of Materials for the Experiments

(Quantities are 1 each except where noted in parenthesis)

A collection of all of the parts listed here (excluding locally available parts, and tools) is available as a kit (#32DBEDFL08) from *Electronix Express* (http://www.elexp.com).

Resistors (1/4 W except where noted)

10 Ω
22 Ω, 2 W
47 Ω
100 Ω(2)
160 Ω
180 Ω
220 Ω
330 Ω (4)
470 Ω
510 Ω
560 Ω
620 Ω
680 Ω
1.0 kΩ (3)
1.2 kΩ
1.5 kΩ
2.0 kΩ
2.2 kΩ (2)
2.7 kΩ
3.3 kΩ
3.9 kΩ
4.7 kΩ (2)
5.1 kΩ
5.6 kΩ
6.2 kΩ
6.8 kΩ
8.2 kΩ (4)
10 kΩ (5)
15 kΩ
20 kΩ
22 kΩ (2)
27 kΩ
33 kΩ
39 kΩ
47 kΩ
56 kΩ
68 kΩ
100 kΩ(3)
150 kΩ
220 kΩ

Resistors (continued)

330 kΩ (2)
360 kΩ
470 kΩ
1.0 MΩ (2)
1.5 MΩ
5.1 MΩ

Potentiometers:

100 Ω
1 kΩ
5 kΩ
10 kΩ
100 kΩ

Capacitors:

100 pF (3)
1000 pF (2)
2200 pF
0.01 μF (4)
0.1 μF (4)
0.22 μF
1.0 μF (4)
10 μF (2)
47 μF
100 μF
220 μF
1000 μF (2)

Inductors

2 μH (wound in class)
25 μH
150 μH
15 mH

Diodes:

Rectifier1N4001 (4)
Signal 1N914 (2)
Zener, 1N4733A
Photo MRD500
LEDs (two red, one green, one yellow)
MV2109 varactor

Transistors:

2N3053 with heat sink
2N3904 npn transistor (3)
2N3906 pnp transistor
2N4123 npn transistor
2N5458 n-channel JFET (2)
J176 p-channel JFET
MPF102 n-channel JFET
MRD300 phototransistor

Integrated Circuits

LM741C op-amp (3)
555 timer
LM565 phase-locked loop
7490A decade counter
7493A 4-bit ripple counter
7805 or 78L05 regulator

Miscellaneous:

2N5060 SCR
2N2646 UJT
Small 8 or 16 Ω speaker
3rd stage IF transformer
   (20 kΩ primary to 5 kΩ
   secondary) Mouser
   42IF303
NTC 5 kΩ thermistor
   (Mouser part number
   527-2006)

Locally obtained parts (not part of the kit from *Electronix Express*)

Heat shrink tubing (for light
   baffle) that fits over
   phototransistor
Light source (bright lamp
   or flashlight)
30 cm twisted-pair wire
Masking tape
Milliammeter I_{FS} = 10 mA
Milliammeter I_{FS} = 20 mA
Small 9 V battery
12.6 V ac center-tapped
   transformer with fused
   line cord
Soldering iron

Appendix B:
Quick-Start Guide for the Programmable Analog Module

Servenger

www.servenger.com

Programmable Analog Module

PAM-5002R Series

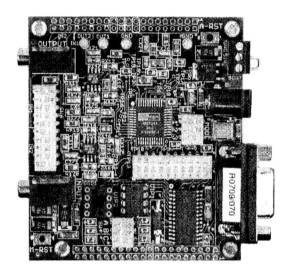

- **Fully supported Anadigmvortex Programmable Analog Signal Processor**
- **6 analog inputs and 3 outputs**
- **On-board +/- 5V power supplies**
- **RS-232 serial interface to host**
- **Operates from host computer or EEPROM**
- **Fully compatible with AnadigmDesigner2 CAD software (available free from Anadigm)**
- **Use for engineering demos or install as a system module in a larger product**
- **Ideal for university level electrical engineering classes and projects**
- **Instructional materials from Servenger website or Pearson/Prentice-Hall**

Quick Start Guide

Obtaining the AnadigmDesigner®2 software and installing on a PC:

To get a Trial copy of the software:

Step 1. Go to **www.anadigm.com** and click on the *FREE SOFTWARE DOWNLOAD* button.

Look for this →

Step 2. A typical web account registration process will start. Enter a UserID and Password to use the website and download AnadigmDesigner®2. Save the account information, complete the registration, download and save the software. Anadigm will email a License ID and License Key.

Step 3. Double click on the saved Setup file to start the installation process. A typical Install Wizard will appear. Use the web account information and **Trial** as the License Key or the full License ID and License Key if available from Anadigm. The Install Wizard will conclude with a check box to **Launch AnadigmDesigner2**. Click **Finish**.

Setting up the Programmable Analog Module (PAM) for AnadigmDesigner2:

Needed:

- AnadigmDesigner2 (referred to as "AD2") computer aided design (CAD) software installed on a Microsoft Windows® based PC.

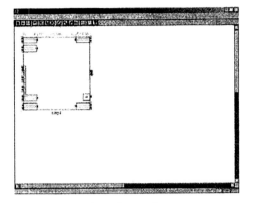

Looks like this when running →

- DB-9 RS-232 Serial Cable (see *Advisories* on last page about USB-to-Serial adapters).
- DC power source: Use either
  - A typical 9VDC output, 1200 mA or larger capacity, AC wall plug power adapter fitted with a 2.1 mm ID x 5.5 mm OD output connector wired as CENTER POSITIVE, or
  - A low noise +5V bench power supply
- Typical set of powered desktop PC stereo speakers with the 3.5 mm stereo plug connector.

Step 1. Connect the DC power source, either using the +9VDC power adapter or a +5VDC bench supply:

    A. Using the supplied +9VDC power adapter per the left side photo below:

        1) Plug the power unit into the AC outlet and connect into the DC Power Jack on the PAM.

        2) Make sure the metal clip connecting 5VOUT to 5VIN is installed (per arrow).

    B. Using a +5VDC bench power supply connected per the right side photo below:

        1) Remove the metal clip connecting 5VOUT to 5VIN (per arrow).

        2) Install GND and +5VDC wire connections to the power terminal per the photo

In both cases the green **+5V ON** LED near the DC Power Jack should be ON continuously.

Step 2. Press the Master Reset pushbutton (labeled M-RST) in the corner of the PCB opposite the power terminal. The two LEDs near the DB-9 connector should flash red and then green in succession.

Step 3. Connect the Serial Cable from the PC to the RS-232 DB-9 port on the PAM

Step 4. Connect the 3.5 mm stereo cable from the PC speakers into the stereo Output jack (labeled **OUTPUT**). Make sure the PC speakers have power and are turned on.

Creating and downloading a sample audio signal source:

Step 1. Every time AD2 starts up a small window will appear asking which Anadigm chip to use. Look at the small lettering in the center of the space above the square design area window which should say AN221E04. If yes then select the **Continue with default chip** button and proceed. If some other part number shows then select the **Choose a different chip** button and change to AN221E04.

Step 2. Point to the pull down menu under **Settings** and select **Preferences**. The Chip Type should be AN221E04 as established above. Select the **Port** tab and then activate the **Select Port** pull down option arrow to show the available options. In general, **COM1** is the best choice but this is PC dependent. Select **COM1** for now, select the **Apply** button and then **OK**. If Step 3 below does not work then select another COM port option and try Step 3 again. Be sure the serial cable is secure.

Step 3. Start AD2 and select **Target > Display Board Information**. A window will appear with the PAM software version and the part number of the Anadigm IC used (AN221E04). Click **OK**.

Step 4. Select **Edit > Insert New CAM**. CAM means "Configurable Analog Module" which is the AD2 name for the pre-defined analog functions. The index to the library of CAMs will be displayed. Point to and double click the Sinewave Oscillator CAM which will appear attached to the tip of the cursor. Drag it into the AD2 design area and click once to drop it.

Step 5. The parameter set up window specific to this CAM will appear. Click **OK** to accept defaults.

Step 6. Select **Settings > Active Chip Settings > Clocks** to adjust the system clock parameters. Point to **Clock 1** setting and use the LEFT/RIGHT slider to set this clock to be divided by **400** times from the system clock frequency to be **40.0 kHz**. Leave all other clock settings unchanged and click **OK**.

Step 7. Point to the Sinewave Oscillator CAM icon, right click and select **CAM Settings**. Set the parameters as follows to achieve **440 Hz (0.44 kHz)** oscillator output:
- Point to **Clock A** and use the pull down option arrow to select **Clock 1** (at **40 kHz**)
- Point to the **Osc. Frequency** value window (temporarily red), enter **0.44** and then **OK**
- Point to the **Peak Amplitude** value window to confirm **3.6 Volts** (default) and then **OK**.

Step 8. Point to the red output port below the Sinewave Oscillator CAM icon and watch the wiring tool pointer appear. Drag the pointer to the upper right output port and release to set the wire. Point to any place along the new blue wire just installed to start a new wire branching off of this wire. Drag the pointer to the lower right output port and release to set the wire. There should now be connections between the output port of the Sinewave Oscillator and the two output ports of the Anadigm IC. The wire just created is labeled "**n1**".

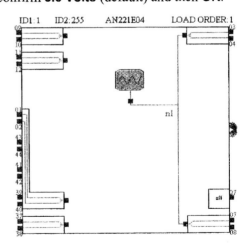

Step 9. Point to the **Download** icon (it has a blue downward pointing arrow). Click once to download the configuration file representing this design into the Anadigm IC on the PAM. The PAM should show three green LEDs as ON with the 440 Hz Tuning-A tone coming out of the speakers. The green LED next to DIP2 socket indicates that a correct Anadigm configuration download has been completed and the analog signal processing is running.

Step 10. The voltage output waveforms can be observed at the terminal posts labeled OUT1 and OUT2.

Storing a copy of the downloaded configuration file in the on-board EEPROM:

As shipped the PAM-5002R will download configuration files directly into the Anadigm IC. To format and store a configuration file into the supplied EEPROM for the Anadigmvortex static configuration mode (MODE = 1) set Switch 4 – 3 (labeled "C") to the ON position. Configuration downloads will proceed as before (a bit slower) but a reformatted version of the configuration file will be put into the SPI serial EEPROM installed in the socket labeled 'DIP1'. To demonstrate the stored contents of the EEPROM:

- Disconnect the power and remove the RS-232 cable
- Relocate the EEPROM to the socket labeled 'DIP2'
- Set all 10 switches in Switch 3 to OFF to isolate the Anadigm IC from the microprocessor
- Set Switch 5 (labeled 'MODE') to OFF – OFF – ON (1 = OFF, 2 = OFF, 3 = ON)
- Reconnect the power.

The configuration file stored in the EEPROM to create the 440 Hz Tuning A will be immediately loaded into the Anadigm IC and the tone will appear at the speakers. This process will repeat after every power up and after the Analog Reset pushbutton is pressed and released.

Advisories about USB-to-Serial adapters, flow control and restarting the AD2:

*#1 – **Use of the Servenger PAM-5002R is not recommended with USB-to-Serial adapters**. However if your PC has only a USB port and no Serial port you may try to use a USB-to-Serial adapter device. No USB-to-serial adapter product is recommended as more likely to succeed than another. An adapter that worked last year with one particular PC is not guaranteed to work purchased new this year nor to work with another PC. Please demonstrate success with your PC before committing to purchase. Selecting between Software and Hardware flow control options per #2 below may be helpful.*

*#2 – As shipped, the PAM-5002R is set to use RS-232 **SOFTWARE** flow control (XON-XOFF command character protocol) to be compatible with most desktop PCs. You can experiment with **HARDWARE** flow control (CTS-RTS signal line handshake) to see if this works better for your equipment by setting Switch 4 – 2 (labeled "B") to ON.*

#3 – If the RS-232 Serial cable is disconnected and then reconnected to the PAM after a configuration download, it may be necessary to save the AD2 design file and then close and restart the AD2 software to restore the logical connection between AD2 and the microprocessor on the PAM.

See the Servenger website at www.servenger.com to download the *Technical User Manual*.

Servenger LLC
**515 NW Saltzman Rd., #904
Portland, Oregon 97229**

Web: www.servenger.com
Email: sales@servenger.com